Studies in Computational Intelligence

Volume 1280

The series "Studies in Computational Intelligence" (SCI) publishes new developments and advances in the various areas of computational intelligence—quickly and with a high quality. The intent is to cover the theory, applications, and design methods of computational intelligence, as embedded in the fields of engineering, computer science, physics and life sciences, as well as the methodologies behind them. The series contains monographs, lecture notes and edited volumes in computational intelligence spanning the areas of neural networks, connectionist systems, genetic algorithms, evolutionary computation, artificial intelligence, cellular automata, self-organizing systems, soft computing, fuzzy systems, and hybrid intelligent systems. Of particular value to both the contributors and the readership are the short publication timeframe and the world-wide distribution, which enable both wide and rapid dissemination of research output.

Indexed by SCOPUS, DBLP, WTI AG (Switzerland), zbMATH, SCImago.

All books published in the series are submitted for consideration in Web of Science.

Roussanka Loukanova
Editor

Computational Linguistics, Information, Reasoning, and AI 2024

Springer

Editor
Roussanka Loukanova
Institute of Mathematics and Informatics
Bulgarian Academy of Sciences
Sofia, Bulgaria

ISSN 1860-949X ISSN 1860-9503 (electronic)
Studies in Computational Intelligence
ISBN 978-3-032-21142-2 ISBN 978-3-032-21143-9 (eBook)
https://doi.org/10.1007/978-3-032-21143-9

This Springer imprint is published by the registered company Springer Nature Switzerland AG
The registered company address is: Gewerbestrasse 11, 6330 Cham, Switzerland

Preface

The chapters in this book evolved from the Special Session on Computational Linguistics, Information, Reasoning, and AI (CLIRAI) (previous alias: CompLingInfoReasAI):

https://www.dcai-conference.net/tracks/special-sessions/clirai

The special session CLIRAI has been part of the International Conference on Distributed Computing and Artificial Intelligence, since DCAI 2017:

https://www.dcai-conference.net

The proceedings of DCAI and its special sessions have been published by Springer Cham:

https://link.springer.com/conference/dcai

Scope

Computational and technological developments that incorporate natural language and reasoning methods are proliferating. Adequate coverage of related works, including advanced applications, encounters difficult problems related to the phenomena of partiality, underspecification, perspectives of agents, and context dependency. These phenomena are signature features of information in nature, natural languages, and reasoning.

The session CLIRAI covers theoretical work, applications, approaches, and techniques for computational models of information, language (artificial, human, or natural in other ways), and reasoning. The goal is to promote computational systems and related models of language, thought, reasoning, and other related processes.

Topics

The special session CLIRAI and this post-proceedings book are dedicated, distinguished forum and publication at which researchers from different subfields of Computational Linguistics present their work. The chapters of this book focus on research relevant to the following topics, across approaches, methods, theories, implementations, and applications:

- Theorem Provers and Assistants
- Model Checkers
- Theories of Computation
- Theories of Information
- Computational Methods of Inferences in Natural Language
- Computational Theories and Systems of Reasoning in Natural Language
- Transfer of reasoning in natural language to theorem provers, or vice versa
- Transfer of reasoning between natural language, theorem provers, model checkers, and various computational assistants
- Translations between natural language of mathematics and formal languages of proof and verification systems
- Controlled Languages of Mathematics
- Computational approaches of Computational Linguistics, e.g., in domain specific areas
- Theories for applications to language, semantics, reasoning, and information processing
- Type Theories for applications to language, semantics, reasoning, and information processing
- Computational Grammar
- Computational Syntax
- Computational Semantics of Natural Language
- Computational Syntax-Semantics Interface
- Interfaces between morphology, lexicon, syntax, semantics, speech, text, pragmatics
- Parsing
- Multilingual Processing
- Large-Scale Grammars of Natural Languages
- Models of computation and algorithms for linguistics, natural language processing, argumentation
- Computational Models of Partiality, Underspecification, and Context-Dependency
- Models of Situations, Contexts, and Agents, for Applications to Computational Linguistics
- Information about Space and Time in Language Models and Processing
- Interdisciplinary Methods
- Integration of formal, computational, model theoretic, graphical, diagrammatic, statistical, and other related methods

- Logic for information extraction or expression in written, spoken, and other modes of language
- Logic for information integrations of diagrams with language
- Computational Models of Argumentations
- Large Language Models (LLM)
- Data Science and Language Processing
- Machine Learning of Language and Reasoning
- Interactive Computation, Reasoning, Argumentation
- Computation with heterogeneous information
- Reasoning with heterogeneous and/or inconsistent information
- Dialog and other Interactions
- Interdisciplinary approaches to language, computation, reasoning, memory
- Computational processing of information and languages in various specific areas and domains, e.g., in forensics, medical sciences, healthcare, jurisdiction, law, etc
- Applications, e.g., to governing, education, business, economy, justice, health, medical sciences, etc
- Computational processing of language based on natural fundamentals of information and languages
- Computational neuroscience of language

Goal and Intended Readers of This Collection of Works

The goal of this collection of works is to promote advanced, intelligent approaches of Mathematical Logic and its applications to systems of proofs and verification, Mathematical and Computational Linguistics, and Natural Language Processing (NLP).

The intended readers of this book are researchers in Mathematical Logic and Computer Science, as well as Mathematical and Computational Linguistics, to advance work on theories and applications to Natural Language Processing (NLP). Artificial Intelligence (AI), Natural Language Processing (NLP), and related subjects.

Sofia, Bulgaria
December 2025

Roussanka Loukanova

Contents

Contributors

Krasimir Angelov University of Gothenburg and Chalmers University of Technology, Gothenburg, Sweden

Wojciech Buszkowski Faculty of Mathematics and Computer Science, Adam Mickiewicz University, Poznań, Poland

Andrea Carrión del Fresno University of Gothenburg and Chalmers University of Technology, Gothenburg, Sweden

Lars Hellan Norwegian University of Science and Technology (NTNU), Trondheim, Norway

Stepan L. Kuznetsov Steklov Mathematical Institute of RAS, Moscow, Russia

Roberto Maieli Department of Mathematics and Physics, Roma TRE University, Rome, Italy

Paweł Płaczek WSB Merito University in Poznań, Poznań, Poland

Aarne Ranta University of Gothenburg and Chalmers University of Technology, Gothenburg, Sweden

Alexey Stukachev Novosibirsk State University, Novosibirsk, Russia; Sobolev Institute of Mathematics, Novosibirsk, Russia

Ekaterina Voloshina University of Gothenburg and Chalmers University of Technology, Gothenburg, Sweden

Ulyana Zaitseva Novosibirsk State University, Novosibirsk, Russia

Natural Language Generation from Wikidata—Architecture, Scalability and Challenges

Krasimir Angelov, Andrea Carrión del Fresno, Ekaterina Voloshina, and Aarne Ranta

Abstract We present an architecture for generation of Wikipedia articles in several languages from Wikidata. The articles cover the topics of countries, cities, people, universities and professions, and vary in size depending on the amount of available information. The architecture is based on the Grammatical Framework, but we made significant extensions to the framework in order to scale it up to the size of Wikidata. We describe the different scalability issues, and we discuss challenges arising from the current Wikidata structure.

Keywords Wikipedia · Wikidata · Grammatical framework · Natural language generation

1 Introduction

Wikipedia is one of the most visited sites on the Web and provides more than 50 million articles in about 300 languages. The coverage of the different language editions is unfortunately unevenly distributed. While the top nine languages comprise more than half of all articles, the bottom half of all languages accounts for less than 10%.

The number of articles does not fully represent the scale of the problem. For example, while for many topics there are extensive articles in English, for other less developed language editions, the corresponding articles are just stubs. Articles in smaller Wikipedia editions are also more likely to be out of date [18]. The opposite

K. Angelov (✉) · A. C. del Fresno · E. Voloshina · A. Ranta
University of Gothenburg and Chalmers University of Technology, Gothenburg, Sweden
e-mail: krasimir@chalmers.se

A. C. del Fresno
e-mail: guscarrian@student.gu.se

E. Voloshina
e-mail: ekaterina.voloshina@chalmers.se

A. Ranta
e-mail: aarne@chalmers.se

R. Loukanova (ed.), *Computational Linguistics, Information, Reasoning, and AI 2024*,
Studies in Computational Intelligence 1280,
https://doi.org/10.1007/978-3-032-21143-9_1

happens as well—small places, local celebrities and customs are best described in the local language and may not appear at all in English.

In 2020, Wikimedia announced the Abstract Wikipedia project which aims to improve the situation by making it possible to generate articles from structured data. If this is possible, then editors will be able to just change the structured data while the text in several languages will be generated automatically in parallel. Moreover, if the text generation is done in a predictable and reliable way, then human post-validation would not be necessary.

We present a prototype[1,2] where we generate articles for countries, cities, people, universities and professions from information in Wikidata. The effectiveness of the method thus naturally depends on the amount and the quality of the data that the Wikidata community has collected. For example, as we will see, the articles for countries and famous peoples can be a page long, while articles for small towns and not so well known people are much shorter. In the extreme, the articles for professions are just a sentence long.

The prototype also relies on the linguistic information in the Resource Grammar Library (RGL) [15] in the Grammatical Framework (GF) [16], the GF WordNet [2, 5], and the entity labels in Wikidata. The use of grammars rather than statistical models guarantees predictability. On the other hand, the task requires wide lexical coverage in the order of millions of entries. This is only achievable by the use of automatic methods. Just as with any automatic method, this is a source of errors, but those can be traced and fixed in a reliable way. In order to handle the large lexicons, we also had to improve the scalability of the framework itself.

2 Related Work

An obvious remedy to the language problem is to use machine translation, and in fact Wikipedia has its own translation tool.[3] The use of the tool, however, has been highly contested since the output from any automatic translation is not reliable and can be used only after post-editing. Unfortunately people forget that and tend to publish unedited content. The tool also does not solve another problem— when the world changes, articles must be updated, and for that we still need human editors even if they do use automatic translation. Generation with Large Language Models from structured data is another approach but it faces the same quality problem [19].

The approach in Abstract Wikipedia is rather generic. It aims to provide a way for programs to read the data and generate text. Whether these programs use grammars or machine learning methods is left to the developers as long as they can reach the quality standards of the community. We see GF simply as one of the possible programming languages for the Abstract Wikipedia.

---

[1] https://cloud.grammaticalframework.org/wordnet/app/wikidata/.

[2] https://github.com/krangelov/gf-wikidata.

[3] https://en.wikipedia.org/wiki/Wikipedia:Content_translation_tool.

Since there are still many options to be explored, the project also encourages the development of alternative approaches. One option is CoSMo [6]—a proposal for a graphical language in the spirit of UML for content selection. Similar to CoSMo, we do consider an extension of the GF language for content selection, but for now, we simply use Python and we leave that language as a future work.

Ninai/Udiron [13] is another system which uses tree editing for content selection, while trying to reuse as much as possible the available information in Wikidata. The tree editing is a technique also available in GF and has been used for a long time [10]. More recently, however, it has been much more popular to use parsing for content creation [1]. People simply find it easier to work with text rather than syntactic or semantic trees. In our vision, content creation should be a combination of light-weight programming in addition to parsing text examples.

A major difference between CoSMo, Ninai/Udiron and our approach is that while the former start from scratch, we reuse and extend an existing system which gives us the advantage of all syntactic and lexical resources created in the past three decades. An early work based on GF [17] showed how these can be used. Here we extend that by integrating more tightly with Wikidata which gives us the best of both worlds.

The main effort in Abstract Wikipedia is now dedicated to Wikifunctions[4]— a wiki of code which allows functions to be edited and executed online. This is also the place where all the language generators will be hosted in the future.

3 Language Resources

3.1 The Resource Grammar Library and GF WordNet

The GF Resource Grammar Library (RGL) [15, 16] is a community project being developed for years, and currently provides the syntax and the morphology for about 40 languages. Unfortunately in terms of lexicon it only includes a small test set with a couple of hundred words.

The lexicon for the project comes from the GF WordNet [2, 5] and contains about 110 000 expressions mostly coming from the Princeton WordNet [9]. We preserve the same semantic relations, but while the original WordNet is composed of lemmas, in GF WordNet the basic unit is an abstract function.

Similar to the concept of a synset, the abstract function represents a particular meaning generalizable across languages. In each language the function is then mapped to an inflection table with a structure compatible with the RGL. Unlike the synset, which can include words synonymous only in a particular context, the abstract functions link words which are translations of each other in as wide a context as possible. In that sense, the abstract function is a concept tighter than a synset and is useful for translation and multilingual generation. Abstract functions with a similar meaning are still grouped in a synset like in Princeton WordNet.

[4] https://www.wikifunctions.org.

The lexicon currently contains 265 languages, but all languages except for English are automatically generated from already existing translation dictionaries and WordNets with the help of automatic disambiguation [5]. Mistakes are thus possible and as part of the current project we also did corrections in the lexicon. Swedish and Bulgarian are notable exceptions which have been extensively checked beforehand [2], and currently more than 50% of the entries are already verified. Another exception is Finnish which was created from the sense-aware translations in FinnWordNet.

Unfortunately, for most of these 265 languages the abstract functions are mapped only to the lemma in each language, while the full inflection tables are available for only 35 languages. We have started, however, the process of automatic induction of morphological rules and currently Albanian, Belarusian, Faroese, Kazakh, Macedonian and Ukrainian are pilot languages. Further investigation into that process is important for covering all languages in Wikipedia, but we leave it as a future work in this paper.

3.2 Integration with Wikidata

During the GF WordNet development, many of the noun synsets were aligned with Wikipedia articles, which helps in disambiguation. When we started working with Wikidata, we merged our alignment with Wikidata's own alignment. In the process many mistakes were identified and fixed on both sides. As a result, currently Wikidata has 30 377 links to Princeton WordNet 3.1. Other lexemes that exist in GF WordNet but not in Princeton WordNet were linked directly with Wikidata entities. We use the resulting links for lexical selection during the generation.

Another missing piece is that while WordNet contains names of people and places, those are limited to important historical figures and well-known places. We filled in the gap by importing names from Wikidata (Table 1).

There are about 7.3 million people and 3.7 million places in Wikidata. Fortunately for the people we do not need to generate a separate entry for every person. In the data the names are decomposed into given name and secondary name. If we only collect those it turns out that by using 64 thousand given names and 531 thousand secondary names, we can describe all the 7.3 million people. For the places, however, we have

Table 1 Lexicon size after extending the grammar with Wikidata names

WordNet	Adjectives, nouns, verbs, etc.	110 thousand	
Wikidata	Given names	64 thousand	Describing
	Family names	531 thousand	7.3 million people
	Place names	3.7 million	
Total		4.3 million	

no other choice but to collect all the names. If we add all these names to the existing WordNet we will end up with a grammar with 4.3 million abstract functions.

Note that although names often propagate with little change from one language to another, there are still differences. This is quite common for place names, but it also happens for people names when two languages use different scripts. Despite these issues, when a name is missing in one language, we still try to borrow it from another related language hoping that this would give us a name close to the native rendering in the original language. For example, a name for Spanish can be borrowed from French and vice versa. For Bulgarian and Russian we also automatically transliterate the names from the Latin script. Borrowing names is obviously not ideal, but it is better than providing no verbalization at all.

4 System Architecture

GF [16] is a programming language specialized in the description of natural languages. A key concept is the abstract syntax—a language independent representation, which is rendered to one or more concrete languages by using grammars.

The abstract syntax is used on two different levels. The Resource Grammar Library applies it only on the level of linguistic concepts (noun phrases, verb phrases, etc). This means that the library alone provides syntactic translations, which are often reasonable, but they can also fail if the proper translation requires different constructions.

In limited domains, we can go further and build a specialized abstract syntax which is more semantic than syntactic, and can thus hide deeper syntactic differences. On that higher level, the RGL syntax is still useful but only as a library which can abstract away low-level syntactic details. In this way, when semantic constructions are implemented on top of the library, in many languages they turn out to be identical. At the same time, when there are exceptions we need to implement only the exceptions while everything else stays shared across languages.

The concept of abstract syntax makes the framework attractive for Wikipedia, but we need to face the challenges. To start with, Wikipedia contains a huge diversity of domains—from geographic places and people to math, physics, biology, etc. We expect each new domain to bring its own concepts but we also expect a lot of overlap.

Wikipedia is also highly dynamic—new content is introduced every minute. Sooner or later the new content brings in new language constructions which also need to be supported in the natural language generation. GF, on the contrary, like most programming languages, is designed to be statically compiled. This means that a change in the grammar requires a recompilation and a restart of the system.

Lastly, the framework has never been used on the scale of Wikipedia. Typical GF applications involve at most thousands of concepts, while here we have millions. The dynamic and diverse nature of the content in Wikipedia and the huge scale, led us to redesign the GF infrastructure itself which we describe in the next section. Here we only focus on how the framework is used in this particular application.

Another issue is the way we typically use the RGL as a library in domain-specific GF applications. Again like in other programming languages, the library is normally statically linked with each application. If we have a lot of domains like in Wikipedia, this means a lot of applications, where each application contains a version of the library.

Here, we solve that by not writing an application grammar but rather to use the library directly from Python. This is a quite unorthodox solution from the GF point of view, but it allows us not to duplicate resources across domains. On the other hand we made the library API directly usable from Python which gives us the same experience.

The architecture for the current prototype is shown on Fig. 1. From the web page the user can query Wikidata for different entities by name. When an entity is selected, the Python program fetches the entity data, potentially together with other related entities. Based on the data, the program plans the document structure and finally produces a sequence of abstract syntax trees. In the process two databases are consulted. The semantic database contains links between the GF WordNet lexicon and the entities in Wikidata. This lets us do the lexical choice when referring to different entities. The second database contains the compiled GF WordNet and the RGL.

The user can do limited content editing directly from the web page. The "Edit" tab on the web page provides an interface where it is possible to click on words and edit their translations and inflection tables. When this happens, a request is sent to the GF compiler which runs on the server. It processes the new definition of the word and stores the compiled version in the database. Changing the natural language generation itself is possible by updating the Python program. We do not allow that from the web yet, since running arbitrary Python code on the server is a risk. The situation will change when we integrate with WikiFunctions which runs Python code in a sandbox.

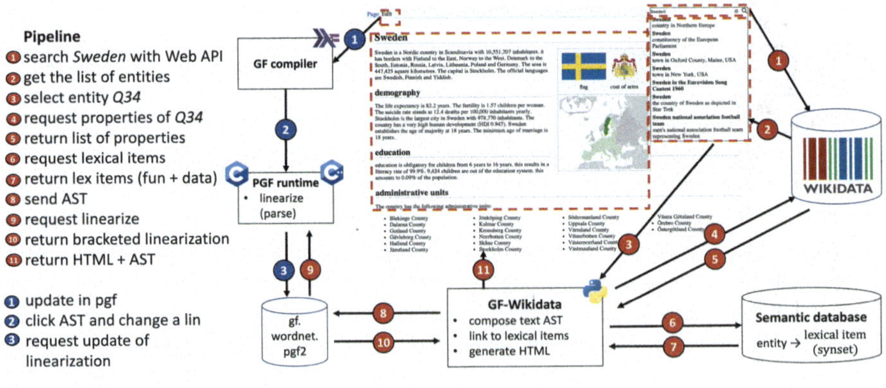

Fig. 1 System architecture

5 Scaling Up

The use of GF on the Wikipedia scale is of course only viable if we can show that the infrastructure can support grammars and lexicons with sizes several orders of magnitude bigger than what we had so far.

The other bottleneck is that currently a single change in the grammar requires a complete rebuild which is not feasible if changes are made all the time. For example, rebuilding the GF WordNet from scratch requires 27 min on a laptop with Intel i7 2.30 GHz CPU and 32 Gb of RAM. The time is lower if the compiler finds that not all source files have been changed, but it is anyway unacceptable if simple changes require minutes of recompilation.

To face the challenges, we have made a major rewrite of both the compiler and the runtime which makes it possible to treat grammars as databases. By default, just like before, the compiler produces a file in the Portable Grammar Format (PGF) [4], which is a low-level and platform-independent representation of the grammar. While loading the PGF, the runtime normally constructs data structures in memory which are then used by the interpreter. The bottleneck here is that the size of the data is limited by the amount of physical memory and swap space. Depending on the size of the grammar the loading may also take significant time. This is noticeable, for example, as a down time when we have to restart our web server.

These problems are alleviated by the introduction of a Native Grammar Format (NGF) [3] which unlike the PGF is different for different CPUs and operating systems. On the other hand, it has the advantage that it can be loaded immediately. For backwards compatibility, by default the format is generated in memory each time a PGF file is loaded. In this mode, the grammar size is still limited by the memory size and the loading may take long time. For bigger files, however, the user can instruct the runtime to store the data in a memory mapped file. In this way, if the grammar needs to be reloaded later, the data is just mapped back to memory. This also avoids the limits of memory and swap space since the file size is only limited by the amount of disk space. When the file starts to significantly exceed the memory size, we of course expect a graceful degradation in speed due to the more frequent swapping between the file and the main memory.

While designing the new format we also took into account the requirement that the grammar must be dynamically modifiable. Unfortunately, this does not come for free. The previous implementation took advantage of the assumption that grammars never change and used that to design more efficient or compact data structures. This is no longer the case. The GF interpreter now provides commands which make it possible to load an already compiled grammar and dynamically change only individual statements. This makes it possible, for instance, to change entries in the lexicon in the matter of milliseconds rather than minutes.

The NGF obviously has the disadvantage that it is platform-dependent. This is especially relevant when changes are made dynamically to an existing grammar. Even if the changes are no longer available as a source code, it is still desirable to be able to move the data from one machine to another. For that purpose, the runtime provides an API to generate an updated PGF file from an existing compiled grammar. That PGF file can then be reloaded on the target machine.

For the actual implementation we first considered using an existing key-value storage such as BerkelyDB, LMDB, LevelDB or the low-level Btree API in SQLite. The problem is that all of these only provide a key-value mapping while for efficiency we also need to store direct pointers from one object to another. In essence we need an Object Oriented storage which is freely available and implemented in a low-level language like C/C++ for compatibility with the rest of the runtime.

Since we did not find anything that is fit for the task, we started building our own database engine inspired by the Lightning Memory-Mapped Database Manager (LMDB) [8]. LMDB is using memory mapped files which speeds up file access by avoiding the intermediate interaction with the swap file. The kernel of the operating system simply loads pages of data on demand directly into the process' memory. Unlike our engine, LMDB is a key-value storage and direct pointers between objects are not supported. Nevertheless, we borrowed the key idea of using memory mapped files but we implemented a persistent heap [7] instead, which stores a graph of objects. We also kept the design of transaction management in LMDB, where copy-on-write is used to guarantee atomicity. The design also allows for readers to work in parallel to the writer. There is still the limitation of a single writer at a time.

We benchmarked the new runtime during the import of names from Wikidata (Table 1). The extraction process traverses a dump of Wikidata and for each name it generates 25 commands in the GF language. The final result is a sequence of ~99 million commands which looks roughly as follows:

```
transaction start

create fun el_salto_PN : PN
create fun kaluga_PN : PN

import -resource alltenses/ParadigmsAfr.gfo
create -lang=Afr lin el_salto_PN = mkPN "El Salto"
create -lang=Afr lin kaluga_PN = mkPN "Kaluga"

transaction commit
```

We start a transaction in the beginning and commit it at the end. For each name we first create one abstract function per name. After that for each language we import a resource module `ParadigmsX` which among other things defines the `mkPN` function. That function takes a string and tries to predict the inflection table for the name. After that we create one lexical entry per name and language.

On a laptop with 32 GB of RAM, an Intel i7 2.30 GHz CPU and a PC711 NVMe SK hynix 1TB solid state drive, the execution of the script finished for ~2.5 h. Given the number of operations we find the time quite satisfactory.

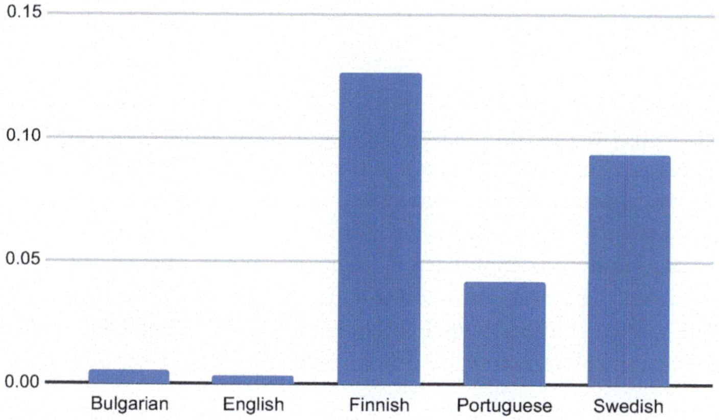

Fig. 2 Linearization time per word in milliseconds for different languages

We also benchmarked the time per word needed to linearize an abstract tree with the new runtime. The results are on Fig. 2. As we can see the time varies per language but never exceeds 0.15 ms per word.

6 English Baseline

For each application domain we first started the development of English. This is not an unrealistic scenario given that the English Wikipedia is also the biggest and most actively developed. Thanks to the multilingual nature of our grammar and lexicon, once we have the English version we also get an out of the box translation for all other languages. We expect something similar to happen in a real deployment. As we will show in the next section, however, adaptations in each language were also necessary.

As a first use case, we chose to generate articles about countries. Later, it was easy to port that implementation to cities since a lot of the properties are shared between countries and cities. The other interesting domain that we explored is the articles for people, which turns out to be particularly diverse since there are more than 7 million people with entities in Wikidata. Finally we looked at universities and professions but as it turns out there was too little data to produce interesting articles in these areas.

6.1 Countries

Wikidata defines 216 countries, and this small set allows us to do extensive testing while the rich information for each of them lets us to generate longer articles. The content is organized into the following sections:
General information: It distinguishes whether a state is a country or an island state and identifies its location in a specific region of the world, such as Scandinavia, the Middle East, Central Asia, Southern Africa, etc. When there is no more specific region we just use the continent. Additionally, this section also covers information regarding the state's borders with neighboring countries, the total area in square kilometers, the capital city and the official and minority languages spoken in the state.

Demography: This section presents quantitative data regarding life expectancy, fertility, suicide rates, and the Human Development Index (HDI). It also includes other demographic details such as the capital city and its population, legal ages of majority and marriage, retirement age, and the official religion of the country[5].

Education: We describe the country's education system, beginning with the age range for compulsory school attendance. The literacy rate is also cited, together with statistics on the number of children who are out of the education system.

Administrative units: This is a simple list of the geographical areas into which the state is divided. The list is useful for its hyperlinks to the articles describing each area.

Politics: Here we include information pertinent to the country's form of government and its current and former head of state and government. In addition, the organizations which the country is currently a member of and/or has belonged to in the past are also mentioned. Further data comprises the democracy index, the degree of civil liberties and political rights as stated by the Freedom in the World report, the designation of a terrorist state, when applicable.

Economy: This section helps to understand the country's economic landscape. We can find economic indicators such as the gross domestic product (GDP), Gini coefficient, inflation rate, total reserve, median income and unemployment rate. The official currency is also introduced, along with the value-added tax (VAT) and the income tax.

Climate: The maximum and/or minimum temperatures recorded in the country.

All that information can also be found in a tabular format in Wikidata, but the textual representation has at least two advantages:

- Related facts are grouped in sections or even in a single sentence. For example, the first sentence for Sweden is:

[5] "Official religion" refers to a religion that is formally recognized by the state. An official religion may not be present in all countries.

> Sweden is a Nordic country in Scandinavia with 10,576,145 inhabitants.

  Here in a single sentence we have packed properties: P361 (part of Nordic countries), P706 (located in Scandinavia), P1082 (population 10,576,145). The textual representation is therefore more structured, compact and easy to comprehend.
- We also actively select which facts are interesting to select. For example, although Wikidata contains information about all previous heads of state and heads of government, we choose to mention only the last two, e.g.:

  > The current head of government is prime minister Ulf Kristersson, who took office after prime minister Magdalena Andersson.

  Listing the succession of prime ministers would only result in a long and boring sequence.
- The entities for countries are linked to separate entities for history, diplomatic relations and economy. Unfortunately, except for the economy all these entities do not bring any new information. Looking at the corresponding Wikipedia pages, however, we see that there is useful information that is still not available in Wikidata. Having that would have been useful to extend the scope of the automatically generated articles. For now we just skip those entities.
- Whenever possible we also add an interpretation. For instance, parameters like the Gini coefficient, the Democracy Index and the HDI are numbers but they also have categorical interpretations which can be turned into textual descriptions.

6.2 Cities

Cities have a lot in common with countries. For instance, they both have area, population, government and economy. On the other hand, features like form of government, HDI, etc., are only relevant on the country level. There are also far many cities than countries. As a consequence, it is not surprising that the amount of data per city is much smaller.

While for countries we mention the general area of the world where the country is, for cities we simply state the country in which it is located. Wikidata also marks specifically the largest cities in every country, as well as the capitals. We use that while introducing the cities. For example:

> Istanbul is the largest city in Turkey with 15,655,924 inhabitants.
> Stockholm is the capital and the largest city in Sweden with 984,748 inhabitants.
> New York is the former capital and the largest city in the United States with 8,804,190 inhabitants.

Here like with countries we introduce the city together with its size in terms of population. After that we follow pretty much the same document plan for countries except that we include only properties relevant to the cities.

One notable problem that we encountered for cities is that Wikidata does not have a consistent taxonomy of settlements. For example, Stockholm is an entity of type *city* and *big city*. At the same time Sofia is not a city but a *metropolis* which is not even a subclass of *city*. Furthermore, small villages in Bulgaria are instances of *"village of Bulgaria"* which is a subclass of *"village"*. On the other hand, smaller settlements in Sweden are instances of *urban area in Sweden* regardless of their sizes.

The diversity of classes makes it more difficult to decide when to use the renderer for cities by just looking at the type of the entity. Moreover the use of one class per country seems completely redundant given that the country is also associated with the settlement by using the property P17 (country).

6.3 People

Since there are more that 7 million people described in Wikidata, we focused our attention to a small list of people shown on Table 2. The set of properties used for describing people is though relatively small and repeated across all the 7 million individuals. The amount of information on the other hand varies a lot. For some

Table 2 Names and the corresponding Wikidata QIDs of the people which we used as inspiration while developing the articles generator

Donald Trump	Q22686	Kamala Harris	Q10853588
Isaac Newton	Q935	Albert Einstein	Q937
Mileva Marić	Q76346	Marie Curie	Q7186
Angela Merkel	Q567	John Lennon	Q1203
Yōko Ono	Q117012	James Hetfield	Q484302
Elvis Presley	Q303	Nelson Mandela	Q8023
Cyril Ramaphosa	Q1148669	Ursula von der Leyen	Q60772
Magdalena Andersson	Q4935873	Sinéad O'Connor	Q193982
Mariah Carey	Q41076	Martin Jacobi	Q122834974
William Chalmers	Q1347373	Aarne Ranta	Q102257069
Per Martin-Löf	Q2067227	Andrey Kolmogorov	Q153224
Mikhail Lomonosov	Q58720	Thierry Coquand	Q3524190
Vladimir Putin	Q7747	Volodymyr Zelenskyi	Q3874799
Benjamin Netanyahu	Q43723	Ismail Haniyeh	Q152378
Greta Thunberg	Q56434717	Xi Jinping	Q15031
Recep Erdoğan	Q39259	Zaha Hadid	Q47780
Gabriel García Márquez	Q5878	Liu Yang	Q235266
Mahatma Gandhi	Q1001	Judith Love Cohen	Q37888317
Frida Kahlo	Q5588	Guido van Rossum	Q30942

people we may only know the name while for others we may produce a full page of text.

We organize the content of the articles in sections, mimicking the ones in the manually created Wikipedia articles:

General information: here in the first sentence, we introduce the person by name and nationality. If we know it, we also include profession, place and date of birth. For some people, the profession is not known and then we need a place holder. If at least the gender is known then we output *man* or *woman*, and if even that is not available then we back it up to *person*. For example:

> Donald Trump is an American moderator, director and president born on 14 June 1946 in New York.
> Mileva Marić was a teacher, physicist and mathematician from Serbia and Germany born on 19 December 1875 in Titel.
> Krasimir Jankov is a Bulgarian man born on 22 March 1986.

After that we continue with potential academic relations, e.g., noted teachers and students associated with the person:

> **John Lennon:** He was a student of Donovan Leitch.
> **Isaac Newton:** His doctoral adviser was Isaac Barrow. He was the supervisor of Roger Cotes, John Flamsteed and William Whiston.

Finally we mention the native language as well as the other languages in which the person is fluent.

Personal life: here we describe the family in which the person is born—mother, father and siblings. After that, marital and extramarital relations as well as children born from different partners.

Personal life is the most complicated section. Some of the people have a lot of relations with children from different partners, which makes it complicated to create a fluent narrative. The document planner looks at the start and end date for different relations and creates a story by trying not to be too repetitive. Furthermore, it takes into account that a relationship may end for different causes. For instance, for Yoko Ono:

> On 20 March 1969, she married British writer John Lennon in Gibraltar.

but she didn't divorce him. Wikidata correctly states the cause for ending the relationship as *death of subject's spouse*.

Education: here we list the educational institutions from which a person has graduated as well as the degree level if available.

Notable works: here we have a bullet list of notable works, i.e., books, films, software, etc. The titles of these works are just taken as labels from Wikidata and are not part of our lexicon. This is possible only because we have a simple bullet list. If we were to incorporate these titles in sentences, we would have to parse them to abstract expressions, which would ensure that we get proper agreements in the sentence.

Awards: here we have two bullet lists - first a list of awards that a person has received, and second all nominations for awards. Like in the previous section, we just use the Wikidata labels of the awards.

6.4 *Universities*

There is not that much information about universities in Wikidata but we included them together with professions for completeness. Every time, an article mentions another entity, the language generator includes a hyperlink to that entity as well. Since the articles for people mention their professions and educational institutions, we wanted to include articles for these as well.

For each university we mention the country in which it is located, and if we know it, the year in which it is established and the person who created it. We decided not to include more precise information, e.g., city or state, since very often the more precise location is also part of the name, i.e., University of Cambridge, University of Gothenburg, Massachusetts Institute of Technology.

The other valuable information is the latest information for number of students, and the current president. The following are some examples:

> Chalmers University of Technology is a university in Sweden established in 1829 by William Chalmers. In 2023 there were 10,773 students. The current president is Martin Jacobi.
> University of Gothenburg is a university in Sweden established on 1 July 1954. There are 37,000 students.
> Cambridge is a university in the United Kingdom established on 1 January 1209. There are 18,977 students.

6.5 *Professions*

The information for professions is even more limited. Professions are described as instances of *job title* and may be subclasses of another profession. In contrast, the corresponding articles in Wikipedia may contain rich and interesting information but that is not available as structured data.

At the moment we just generate a single sentence stating that a profession is a kind of another profession, e.g.:

> A president is a director.

7 Language Adaptations

As we move from the English baseline to different languages, we encounter new linguistic issues. We focused on improving the generation of articles for countries and people in Bulgarian, French, Russian, Spanish and Swedish. The articles for cities does not bring new constructions, and those for universities and professions are trivial.

7.1 Countries

The first and most obvious issue is the need to handle the translation of multi-word expressions—fixed expressions that cannot be split into separate words without altering the meaning or the grammatical structure of the sentence [11]. This characteristic makes the translation task particularly complex, as they often do not translate compositionally. That is, a direct translation often fails to capture the intended concept accurately in its original language. An example is the compound *"life expectancy"* and its Spanish counterpart *"esperanza de vida"*, which literally translates to *"hope of life"*.

The expression 'square kilometres' on Table 3 is another example. In Swedish it is clearly a compound noun—'kvadratkilometer' , while in English it is ambiguous since WordNet describes 'square' as both an adjective and a noun. In French and Spanish the default translation of a compound noun involves the preposition de—'kilomètres de carré' / 'kilómetros de cuadrado' which is clearly not what we need. On the other hand, using an adjectival modification renders the correct expressions as in the table. The Bulgarian and Russian grammars are more flexible and can translate compound nouns by either replacing the modifying noun with an adjective or by inserting a preposition similar to French and Spanish. This means that regardless of whether the expression is treated as a compound noun or an adjectival modification on the abstract level, we always get the right construction in the concrete language.

Table 3 shows one more difference. When expressing a characteristic or an attribute associated with a particular subject in Spanish and French, we often see the preposition 'de' ('of') following the verb 'es'/'est' ('to be'). On the other hand, the preposition is omitted in the other languages, and the copular verb is directly followed by the attribute.

Generally, prepositions do not translate directly, as they may carry different meanings. For instance, when describing the geographical location of Canada, one might

Table 3 An illustrative example of linguistic differences between languages

BUL	Площта е 9,984,670 квадратни километра
ENG	The area is 9,984,670 km^2
FRE	La superficie est de 9,984,670 kilomètres carrés
RUS	Площадь– 9,984,670 квадратных километров
SPA	La superficie es de 9,984,670 kilómetros cuadrados
SWE	Ytan är 9984670 kvadratkilometer

say *"Canada is a country in North America"*. The French translation, however, is *"le Canada est un pays d'Amérique du Nord"*, where the appropriate preposition 'de' (contracted as d') is more similar to the English 'of'.

At the syntactic level, when we want the focus of a sentence to fall on the action over the subject, the passive voice is often used in English. However, there are other grammatical constructions that are preferred in other languages and that are used to express the same idea. That is, the case of the pronoun 'se' in Spanish, utilized with transitive verbs in the active voice to indicate that the interpretation of the verb is passive and that the subject is the recipient of the action. This use of 'se' is a form of the passive voice in Spanish, distinct from the traditional passive voice in English ('to be' + past participle). Similar was encountered in Bulgarian and Russian: in some constructions reflexive forms of verbs are used instead of passive voice forms.

Normally these differences in GF are handled in application specific grammars, since covering all differences in the RGL once and for all is too difficult. Since we decided not to use an application grammar, sometimes we simply have to generate different RGL representations from the Python program for each language.

When the differences are more localized we can instead choose to add new multi-word expressions in the lexicon. We used that for "life expectancy" for instance. On the other hand, constructing expressions by combining existing words is a more light-weight approach if we can still get the right wording in all languages.

In total we added 10 new expressions in the lexicon, and handled 12 other constructions by altering the generation code in Python.

7.2 People

A notable issue in the people's articles is the use of grammatical tense. In English, the past tense is typically used when referring to deceased individuals, while the present tense is reserved for those still alive. That is not the case in Bulgarian and Russian, where the present tense is also used for historical events, e.g., Isaac Newton is still a physicist. To some extent, this applies to Spanish as well, except that Spanish switches to the past tense when an event has clear temporal boundaries, e.g.:

> Sus supervisores doctorales fueron Alfredo Kleiner, Heinrich Burkhardt y Heinrich Weber.

Spanish presents an additional complexity with the distinction between the preterite (pretérito perfecto simple) and the imperfect (pretérito imperfecto), a contrast that does not have a direct equivalent in English. French, on the other hand, behaves more like English but sometimes uses different anteriority, e.g.:

> Il avait obtenu son diplôme à l'École Polytechnique Fédérale de Zurich.

vs

> Il obtenait son diplôme à l'École Polytechnique Fédérale de Zurich.

Gender is another issue to be considered. Obviously, we have the choice between 'he' and 'she' when referring to the person, but there are other aspects as well. First of all, many languages have different nouns for occupations depending on gender. English exhibits some of them (waiter/waitress), but in Romance and Slavic languages this is a lot more common. These distinctions are not well represented in the Princeton WordNet, but over the course of the project, we added many of them to the GF WordNet. How often the female variants of the nouns are used is also language specific. For example, although Russian has many of these, in official text, the male variants are allowed and even preferred even for women. This had to be taken into account in the document generation.

Gender variation shows up also when talking about getting into marriage, since Bulgarian and Russian use different verbs when man or woman gets married.

The different use for passive constructions shows up for people as well as for the countries. When talking about someone being born, English and Bulgarian use passive, while the other languages use a simple intransitive verb – nació, *was born*, е роден, родился

8 Evaluation

One of the things that we want to evaluate is to what extent the GF WordNet lexicon is appropriate as a baseline for generating multilingual content. We always start the generation with one language—English in this case. Thanks to the language independent API of the RGL this automatically produces an out-of-the-box draft for all the 28 currently supported languages. The first step for every language is then to go through the articles and check whether we use the correct lexemes. As it turns out most of the words are used correctly but it may happen that they still have uncertain status in the lexicon since they were never checked before. In those cases we just confirm that the word choice for a given sense is correct. If the word choice is wrong or missing then we simply update the lexicon. In either case the validation status or the potentially new word choice is saved in a database and preserved for future applications.

After all words are validated, we get an improved draft, which is almost fluent but some problems still remain because one and the same linguistic construction can be used in different ways in different languages. Finally, we fix non-compositional constructions which results in a final publishing-quality article.

To evaluate the effect of each step, we compute the BLEU scores [14] for each of the drafts against the final article. Traditionally, the BLEU score is used to evaluate a machine translation against a human translated document. Here we use it in reverse, i.e., we work on the language generator until we get the final article and then we use that instead of the gold standard. This means that for the final article we always get a score of 100 for all languages which would be very hard in machine translation (Table 4).

Table 5 shows the evaluation scores for the articles about countries. What we

Table 4 BLEU scores for the generated articles about countries after each phase

	Initial draft				Improved draft				Google translate			
	1-gram	2-gram	3-gram	4-gram	1-gram	2-gram	3-gram	4-gram	1-gram	2-gram	3-gram	4-gram
Bulgarian	80.04	72.94	67.05	**61.12**	99.17	98.88	98.65	**98.41**	41.53	34.38	30.01	**26.31**
English	94.08	93.13	92.19	**91.19**	100.00	100.00	100.00	**100.00**	–	–	–	**–**
French	64.43	53.94	44.80	**38.01**	95.60	93.45	91.10	**88.75**	76.17	69.32	63.98	**59.99**
Russian	43.21	26.01	17.28	**11.77**	93.12	88.82	85.68	**83.28**	63.75	51.60	42.63	**36.07**
Spanish	73.57	62.48	52.82	**44.75**	96.13	93.75	91.10	**88.31**	82.62	72.99	64.77	**58.26**
Swedish	81.82	76.44	71.67	**67.01**	99.26	98.94	98.75	**98.57**	67.07	56.51	49.23	**43.76**

Table 5 BLEU scores for the generated articles about people

	1-gram	2-gram	3-gram	4-gram
Bulgarian	85.70	83.98	83.16	82.80
French	87.38	82.75	78.81	75.31
Russian	85.26	80.44	76.22	72.45
Spanish	86.27	83.82	81.71	79.89

observe is that the BLEU scores for the initial draft against the final version varies from 38 to 67 with two exceptions. English is our source language and there we only fixed the capitalizations of some words. This explains the high-scores. The other exception is Russian, where the grammar was recently rewritten from scratch, and during the project we found both incorrect lexemes as well as mistakes in the syntactic rules. As a result the initial scores are much lower than for the other languages.

These scores basically measure how good is the current RGL in combination with the GF WordNet with no additional work. When BLEU is used for machine translation, scores above 50 generally reflect good and fluent translation [12]. Indeed, even without any improvements we found the articles to be quite understandable but still far from production quality.

After fixing wrong word translations or filling in missing ones, we get the improved draft, and there the scores already go above 80. For translation, BLEU scores above 80 are very difficult to obtain because a single sentence can be translated in many possible ways. Since our generation method is predictable, we can control how our output should look like. In that sense it is not surprising that we can get scores up to 100. It is interesting that the effect of non-compositional translations is only 2-17 BLEU points. The difference is of course language dependent but it shows that we can get quite good translations almost for free for many languages.

For comparison we also translated the final versions of the English articles with Google Translate to all the other languages and analyzed the results. As expected the BLEU scores are much lower, but comparing only the numbers would be unfair since Google Translate may also choose to translate in other ways which are also correct. Below, we only highlight cases where the translation is clearly wrong.

Occasionally, it appears that Google's translation of a term may not be appropriate for the context of the sentence, resulting in the incorrect sense of the word. When translating the sentence "the fertility is X children per woman", the term "fertility" is understood as the ratio of live births in an area[6]. In Google's Spanish translation we find the term "fertilidad" ('fertility'), which pertains to the state of being fertile or capable of producing offspring[6]. Hence, this interpretation does not align with the intended meaning in the sentence. The same is also mirrored in the translations of the other languages.

Another notable difference is the inaccuracies or omissions that can occur when translating information from the original text. For example, in a sentence describing the official and unofficial languages spoken in a country. Google's translation omits the unofficial languages by stating that many other languages are spoken in the country, and only highlighting the official ones. The result is a translation that does not convey the same meaning as the original text. It appears that long sentences may contribute to the occurrence of hallucinations, as this behavior is not seen in similar but shorter sentences.

In addition, Google's version fails to provide a consistent translation of some acronyms or organization/institution names that have their equivalence in other languages. An example includes the IFC (International Finance Corporation) or the Commonwealth of Nations, which are presented as is, without providing their Spanish equivalents, 'CFI' and 'Mancomunidad de Naciones' respectively.

Finally, several proper names are not capitalized in the translated version. Even though this matter may not directly concern the translation task, it does impact the overall quality of the output text.

While for countries we made an extensive evaluation, for people we kept things simple. We took the articles produced for the people on Table 2 and we manually post-edited the results. Table 5 shows the BLEU scores between the currently generated articles and the manually edited ones. A manual analysis shows that the main source of errors is the injection of names from different languages in the current language. For example, names of awards in the Bulgarian articles, frequently show up in Russian or Serbian. While this choice still introduces lower character-based error rate, the mix of languages makes the output look less professional. On the other hand, while post-editing the text, it is still worthwhile to start with a name in a related language rather than with English.

9 Challenges with Wikidata

While working on the project we encountered a number of challenges related to the current state of Wikidata. These are summarized in this section.

One recurring issue is the way we refer to people. Normally we will use properties P735 (given name) and P734 (family name) to extract the corresponding lexemes which we imported from Wikidata to the GF WordNet. With these two lexemes we will construct the full name. The problem is that for 11% of the people, the first name is missing and for 39% the family name is missing. In these cases, as a backup, we use only the first or the second name. This, however, is far from ideal. For example, as it happens, while reading the article about Andrey Kolmogorov we see that, among other people, he supervised someone called *Aleksandr*. Following the link, we also read for Aleksandr that:

> He is the supervisor of Anna, Denis, Maxim, Pavel and Maxim.

By inspecting the Wikidata entity, we see that this is the mathematician Alexander Bulinskiy, who does not have a Wikipedia article in English but is described in the Russian, Bashkir and Arabic Wikipedias.

A related issue is that there are people with several given or family names, or with no family name at all. In these cases it is hard to know what is the best way to refer to a person, without a further analysis. For example, the current article for Sweden mentions king Carl as a head of state, but he is better known as Carl XVI Gustaf. In fact Carl XVI Gustaf has two more given names and no family name.

Solving these issues would require an offline analysis of all Wikidata labels for people, and the extraction of given names and family names. Since the given and family names are entities on their own, sometimes new entities need to be created and then linked to the right people. This is not a completely strightforward task. In some cultures, the given name comes first, in others second. Sometimes there are regnal ordinals and sometimes prepositions like 'von'. In cases where the name consists of several words, it might be ambiguous which are the given and which are the family names. There are also labels like 'Eleanor of Naples, Duchess of Ferrara' which clearly contain more than just the name. In cases, where a person is best known with a certain name, it should also be made easy to automatically recover that.

The labels for geographical objects (countries, cities, etc.) are not made to be used for natural language generation either. For example, the label 'Sommersdorf, Mecklenburg-Vorpommern' is very descriptive but if taken literally would lead to sentences like:

Sommersdorf, Mecklenburg-Vorpommern is a municipality in Mecklenburg-Vorpommern

The qualification in the label is not strictly necessary since the entity is also described by property P131 as belonging to Mecklenburgische Seenplatte.

Moreover, some names may have definite articles in specific languages— the Alps, the Netherlands. This is further complicated by the fact that whether the article should be used also depends on the context in which the name appears. Different geographical locations are also used with different prepositions—in the Alps, on the Balkans at the South Pole. The choice of preposition also depends on the language. Currently we have implemented the articles and the prepositions for geographical names that are also included in the Princeton WordNet, but naturally there are many more examples in Wikidata. Identifying them all is a long process which must be supported by properties in Wikidata.

Generally names must be lexemes as well. For instance, they might inflect in different grammatical cases, and they might have grammatical gender. Wikidata already allows the storage of lexemes but this is not fully used yet.

Another technical limitation is the access to Wikidata. While we have worked to optimize our natural language pipeline, the remaining limitation is still that every request to Wikidata takes about half a second. We tried to bundle requests in batches but the delay is still noticeable since the article generation hands for a while, every time when we have to retrieve a bunch of entities.

Note that we use the JSON Web API which given a QID retrieves an entire entity. In our experiments, using the SPARQL entry points turns out to be even less efficient, and the request may even be denied after a certain time out limit.

10 Conclusion

We find the quality of the generated articles satisfactory and the generation process is easy to steer in the desired direction. Some issues with names of people, places, universities and awards, however, have been mentioned. Since we have control over the process there is no risk of hallucinations unlike with Large Language Models.

The lexicon is largely automatically generated and thus there are mistakes, but the same is true for any machine learning method. On the other hand, we make it possible to manually correct the lexicon, which means that it will improve over time. Because of this, community involvement is crucial, if any natural language generation system is to be deployed in practice.

In terms of performance, the bottleneck is in the communication with Wikidata. Because of that, the user cannot see the requested page immediately. We know that the Wikifunctions team have discussed plans for additional caching which might improve the situation.

Using Python in combination with GF works, but a special purpose programming language might be better. Currently the Python code has to do a lot of pattern matching to find which combination of properties exists for the current entity and how to package them in compact sentences. This can be done more ergonomically in a specialized language.

References

1. Angelov K (2009) Incremental parsing with parallel multiple context-free grammars. In: European chapter of the association for computational linguistics (2009)
2. Angelov K (2020) A parallel WordNet for English, Swedish and Bulgarian. In: Proceedings of the 12th language resources and evaluation conference. European Language Resources Association, Marseille, France, pp 3008–3015https://www.aclweb.org/anthology/2020.lrec-1.368
3. Angelov, K (2024) A scalable database for syntactic and lexical resources. In: CLIRAI 2024: computational linguistics, information, reasoning, and AI
4. Angelov K, Bringert B, Ranta A (2010) PGF: a portable run-time format for type-theoretical grammars. J Logic Lang Inform 19:201–228
5. Angelov K, Lobanov G (2016) Predicting translation equivalents in linked wordnets. In: The 26th international conference on computational linguistics (COLING 2016), p 26
6. Arrieta K, Fillottrani P, Keet M (2024) CoSMo: a multilingual modular language for content selection modelling. In: Symposium on applied computing
7. Atkinson M, Chisholm K, Cockshott P, Marshall R (1983) Algorithms for a persistent heap. Softw Pract Exp 13:259–271. https://doi.org/10.1002/spe.4380130306
8. Chu H (2011) MDB: a memory-mapped database and backend for OpenLDAP. Symas Corporation, OpenLDAP Project. https://www.openldap.org/pub/hyc/mdb-paper.pdf
9. Fellbaum C (1998) WordNet: an electronic lexical database. MIT Press (1998)
10. Hallgren T, Ranta A (2000) An extensible proof text editor. In: Parigot M, Voronkov A (eds) Logic for programming and automated reasoning. Springer, Berlin Heidelberg, Berlin, Heidelberg, pp 70–84
11. Huddleston R, Pullum GK (2002) The Cambridge grammar of the English language. Cambridge University Press

12. Lavie A (2011) Evaluating the output of machine translation systems. In: Proceedings of machine translation summit XIII: tutorial abstracts. Xiamen, China. https://aclanthology.org/2011.mtsummit-tutorials.3
13. Morshed M (2023) Using Wikidata lexemes and items to generate text from abstract representations. Sem Web J 3564–4778
14. Papineni K, Roukos S, Ward T, Zhu WJ (2002) BLEU: a method for automatic evaluation of machine translation. In: ACL, pp 311–318
15. Ranta A (2009) The GF resource grammar library. Linguistic issues in language technology
16. Ranta A (2011) Grammatical framework: programming with multilingual grammars. CSLI Publications, Stanford. iSBN-10: 1-57586-626-9 (Paper), 1-57586-627-7 (Cloth)
17. Ranta A (2023) Multilingual text generation for abstract Wikipedia in grammatical framework: prospects and challenges. Springer, Cham, pp 125–149
18. Vrandečić D (2020) Architecture for a multilingual Wikipedia
19. Zhang Y, Li Y, Cui L, Cai D, Liu L, Fu T, Huang X, Zhao E, Zhang Y, Chen Y, Wang L, Luu AT, Bi W, Shi F, Shi S (2023) Siren's song in the AI ocean: a survey on hallucination in large language models

A Survey on Language Completeness of the Lambek Calculus and Its Extensions

Stepan L. Kuznetsov

Abstract The most natural class of models for the Lambek calculus, according to its linguistic applications, is formed by models on algebras of formal languages, or L-models for short. While the Lambek calculus is known to be complete w.r.t. this class of models (Pentus 1995), its natural extensions are often incomplete. We give a survey of known (in)completeness results and present a couple of new ones.

Keywords Lambek calculus · Completeness · Formal languages

1 Introduction

The *Lambek calculus* [27] is a non-commutative substructural logic used for modelling natural language syntax via the formalism of categorial grammars. Formulae of the Lambek calculus, being built from "propositional" variables using two directed implications (called divisions, or residuals) and product (a kind of conjunction), are called *syntactic types* and represent syntactic roles of words and subphrases within a phrase. This suggests, as the most natural one, the interpretation of Lambek formulae as *formal languages* over an alphabet Σ. Such models of the Lambek calculus are called *formal language models,* or *L-models* for short.

The seminal result of Pentus [33] establishes *completeness* of the Lambek calculus w.r.t. L-models. Pentus' theorem designates the Lambek calculus among other calculi used in categorial grammars (others include, e.g., basic categorial grammars, combinatory categorial grammars etc.) as *the* logic of language concatenation and residuation. Namely, natural principles like transitivity: $A / B, B / C \to A / C$, type lifting: $A \to B / (A \setminus B)$, or currying: $(A \cdot B) \setminus C \to B \setminus (A \setminus C)$, which are used in categorial grammar parsing, need not to be postulated on their own, but are rather *derivable* in the Lambek calculus. (And those which are not derivable, are not generally true in L-models, thus, they are not valid principles.)

S. L. Kuznetsov (✉)
Steklov Mathematical Institute of RAS, Moscow, Russia
e-mail: sk@mi-ras.ru

R. Loukanova (ed.), *Computational Linguistics, Information, Reasoning, and AI 2024*, Studies in Computational Intelligence 1280,
https://doi.org/10.1007/978-3-032-21143-9_2

Extending the Lambek calculus with other natural operations on formal languages, however, may destroy completeness w.r.t. L-models. Issues with completeness are typically connected to the distributivity law for set-theoretic operations (intersection and union), the hidden union inside the Kleene star (in combination with intersection), and specific properties of constants zero and one, which denote, resp., the empty language and the singleton of the empty word.

Moreover, one should distinguish *weak* and *strong* competeness. Pentus' theorem discussed above is a weak completeness result: if a principle (*sequent*) of the form $A \to B$ is generally valid in L-models, then it is provable in the Lambek calculus. In contrast, talking of reasoning from (even finite) set of hypotheses, derivability in the Lambek calculus is not the same as semantic entailment on the class of all L-models. Namely, there exist two sequents, $C \to D$ and $A \to B$, such that the latter is true in all L-models where so is the former; however, $A \to B$ is not derivable in the Lambek calculus extended with $C \to D$ as a hypothesis. This is the lack of strong completeness.

Most of these issues can be resolved by modifying the interpretation of Lambek operations. Usually this is achieved by adding a closure operation on the languages. One of such approaches is based on *syntactic concept lattices* (SCL) suggested by Clark [10] and Wurm [37]. In this modification of L-models, called SCL-models, each language M is extended by adding all the words which may appear in the same *contexts* as words from M. More formally, the closure of M is $\{v \mid \forall u, w\ ((\forall v' \in M\ (uv'w \in L_0)) \Rightarrow uvw \in L_0)\}$, where L_0 is a fixed language. The SCL consists only of closed languages (i.e. those which coincide with their closure). Some of the operations do not keep the languages closed.[1] In SCL-models, after performing such an operation, closure is applied.

Most extensions of the Lambek calculus (a notable exception being the one via constant zero) are strongly complete w.r.t. SCL-models [24, 37]. Throughout this paper, however, we consider L-models in the classical sense, and discuss where completeness holds, and where it fails. This paper is designed as a survey, so most of the theorems are previously known ones. Our purpose is to conveniently place them together for usage and reference. There are, however, some minor new results presented here, in order to naturally fill some gaps in the previous knowledge. This includes the lack of compactness for reasoning from hypotheses in the presence of product (Prop. 2) and the usage of intersection with constant one as a substitute for the exponential modality, which gives an alternative proof of undecidability in the presence of the unit constant (Sect. 3 of [16]).

[1] For example, if $L_0 = \{c_1ad_1, c_2bd_2, c_3abd_3, c_3ed_3\}$, then $\{a\}$ and $\{b\}$ are closed, while the closure of their product $\{ab\}$ is $\{ab, e\} \neq \{ab\}$. Similar examples exist for union and Kleene star. In contrast, intersection and Lambek division of closed languages is also a closed language.

2 The Lambek Calculus and L-Models

Let us recall the formulation of the Lambek calculus. Let us fix a finite set of variables (primitive types) Var, denoted by p, q, r, s, maybe with subscripts. Lambek formulae, denoted by capital Latin letters, are built from variables using three binary operations: $\backslash$ (left division), $/$ (right division), and $\cdot$ (product). The set of formulae is denoted by Fm; in what follows, we shall tacitly extend formulae (and, resp., the set Fm) with more operations.

The Lambek calculus is traditionally formulated as a sequent system. (Another, equivalent, version is a natural deduction calculus—see, e.g., [29, Sect. 2.2].) Capital Greek letters denote finite sequences of formulae (including the empty one), i.e., elements of Fm^*. *Sequents* are expressions of the form $\Pi \to A$. The left-hand side Π is the *antecedent* and the right-hand side A is the *succedent* of $\Pi \to A$.

The Lambek calculus comes in two variants, depending on whether *empty antecedents* are allowed. In the original Lambek calculus, all sequents were required to have non-empty antecedents. This constraint, sometimes called *Lambek's non-emptiness restriction,* has linguistic motivations. Namely, sequents with empty antecedents allow assigning certain syntactic types to empty strings, which leads to recognising ungrammatical phrases as valid ones [29, Sect. 2.5]. In particular, the empty string could serve as an "adjective," which validates, along with "*very interesting book*" also the incorrect phrase "*very book.*"

Let us start with semantics and define L-models for the Lambek calculus obeying Lambek's non-emptiness restriction.

Definition 1 An *L-model* is a pair $\mathcal{M} = (\Sigma, \mathsf{w})$, where Σ is an alphabet (finite non-empty set) and $\mathsf{w} \colon \mathrm{Fm} \to \mathcal{P}(\Sigma^+)$ is a *valuation function* which maps Lambek formulae to formal languages over Σ without the empty word and commutes with the operations in the following way:

$$\begin{aligned}
\mathsf{w}(A \cdot B) &= \mathsf{w}(A) \cdot \mathsf{w}(B) = \{uv \mid u \in \mathsf{w}(A) \text{ and } v \in \mathsf{w}(B)\},\\
\mathsf{w}(A \setminus B) &= \mathsf{w}(A) \setminus \mathsf{w}(B) = \{u \in \Sigma^+ \mid (\forall v \in \mathsf{w}(A))\ vu \in \mathsf{w}(B)\},\\
\mathsf{w}(B \mathbin{/} A) &= \mathsf{w}(B) \mathbin{/} \mathsf{w}(A) = \{u \in \Sigma^+ \mid (\forall v \in \mathsf{w}(A))\ uv \in \mathsf{w}(B)\}.
\end{aligned}$$

A sequent $A_1, \dots, A_n \to B$ (where $n \geq 1$: empty antecedents are not allowed) is true in model $\mathcal{M}$ (notation: $\mathcal{M} \vDash A_1, \dots, A_n \to B$), iff

$$\mathsf{w}(A_1 \cdot \ldots \cdot A_n) \subseteq \mathsf{w}(B).$$

The case without Lambek's non-emptiness restriction is handled by changing Σ^+ to Σ^*, which is the set of all words, including the empty word ε. In order to avoid naming collisions, we call the modified class of models "Lε-models" (cf. [16]).

Definition 2 An Lε-model is a pair $\mathcal{M} = (\Sigma, \mathsf{w})$, where Σ is an alphabet and $\mathsf{w} \colon \mathrm{Fm} \to \mathcal{P}(\Sigma^*)$ is a valuation function obeying the following conditions:

$$\mathsf{w}(A \cdot B) = \{uv \mid u \in \mathsf{w}(A) \text{ and } v \in \mathsf{w}(B)\},$$
$$\mathsf{w}(A \setminus B) = \{u \in \Sigma^* \mid (\forall v \in \mathsf{w}(A))\ vu \in \mathsf{w}(B)\},$$
$$\mathsf{w}(B \,/\, A) = \{u \in \Sigma^* \mid (\forall v \in \mathsf{w}(A))\ uv \in \mathsf{w}(B)\}.$$

We have $\mathcal{M} \vDash A_1, \ldots, A_n \to B$ iff $\mathsf{w}(A_1 \cdot \ldots \cdot A_n) \subseteq \mathsf{w}(B)$, and $\mathcal{M} \vDash {\to} B$ (for the empty antecedent case) iff $\varepsilon \in \mathsf{w}(B)$.

Notice the subtle difference in the two definitions of division operations. Even when both $\mathsf{w}(A)$ and $\mathsf{w}(B)$ do not include ε, in an Lε-model it could still be the case that $\varepsilon \in \mathsf{w}(A \setminus B)$, while in L-models this is impossible by definition. Thus, there exist sequents with non-empty antecedents, which are true in all Lε-models, but not in all L-models. Examples of such sequents include $q \to q \cdot (p \setminus p)$ and $(p \setminus p) \setminus q \to q$.

In what follows, we shall define two versions of the Lambek calculus, with and without Lambek's non-emptiness restriction, and the latter will be a (non-conservative) extension of the former.

The original Lambek calculus, with Lambek's non-emptiness restriction, is denoted by **L**. For the version which allows empty antecedents we shall use the notation $\mathbf{L}^{\varepsilon}$. (The more common notation $\mathbf{L}^*$ would lead to notation clash with Kleene star, see below.)

The axioms and rules of **L** and $\mathbf{L}^{\varepsilon}$ are the same in formulation, as presented below, but in **L** all sequents in a derivation are required to have non-empty antecedents.

$$\frac{}{A \to A}\ Id \qquad \frac{\Gamma, A, B, \Delta \to C}{\Gamma, A \cdot B, \Delta \to C}\ {\cdot}L \qquad \frac{\Pi \to A \quad \Delta \to B}{\Pi, \Delta \to A \cdot B}\ {\cdot}R$$

$$\frac{\Pi \to A \quad \Gamma, B, \Delta \to C}{\Gamma, \Pi, A \setminus B, \Delta \to C}\ {\setminus}L \qquad \frac{A, \Pi \to B}{\Pi \to A \setminus B}\ {\setminus}R$$

$$\frac{\Pi \to A \quad \Gamma, B, \Delta \to C}{\Gamma, B \,/\, A, \Pi, \Delta \to C}\ /L \qquad \frac{\Pi, A \to B}{\Pi \to B \,/\, A}\ /R$$

$$\frac{\Pi \to A \quad \Gamma, A, \Delta \to C}{\Gamma, \Pi, \Delta \to C}\ Cut$$

For "pure" derivability (theoremhood), without hypotheses, the Cut rule is eliminable [27]. Moreover, in this case the only two rules which can make the antecedent empty are $\setminus R$ and $/R$. Thus, instead of imposing a global antecedent non-emptiness constraint, the requirement "Π is non-empty" is added to these two rules. For derivability from hypotheses (i.e., sequents from a given set which are used as if they were extra axioms), it is additionally required that all hypotheses have non-empty antecedents.

Pentus proved completeness of both $\mathbf{L}$ and $\mathbf{L}^\varepsilon$ w.r.t. L-models and Lε-models, respectively.

Theorem 1 ([33, 34]) *A sequent (with a non-empty antecedent) is true in all L-models if and only if it is derivable in* $\mathbf{L}$. *A sequent (possibly with an empty antecedent) is true in all Lε-models if and only if it is derivable in* $\mathbf{L}^\varepsilon$.

Notice that these are *weak* completeness results, as they do not deal with derivability from sets of hypotheses.

In order to discuss *strong* completeness, let us recall the basic definition of semantic entailment. This definition is very general, and deals with abstract logical systems and classes of models.

Definition 3 Let $\mathcal{L}$ be a logic and let $\mathcal{K}$ be a class of models (for the language of $\mathcal{L}$). A set of sequents $\mathcal{H}$ *semantically entails* a sequent $\Pi \to A$ on the class $\mathcal{K}$ (notation: $\mathcal{H} \models_{\mathcal{K}} \Pi \to A$), if for any model $\mathcal{M} \in \mathcal{K}$, in which all sequents from $\mathcal{H}$ are true, $\Pi \to A$ is also true.

By $\mathcal{H} \vdash_{\mathcal{L}} \Pi \to A$ we denote the fact that $\Pi \to A$ is derivable from $\mathcal{H}$ in $\mathcal{L}$.

Definition 4 A logic $\mathcal{L}$ is *strongly sound and complete* w.r.t. the class of models $\mathcal{K}$ if for any sequent $\Pi \to A$ and for any (possibly infinite) set of sequents $\mathcal{H}$ we have

$$\mathcal{H} \vdash_{\mathcal{L}} \Pi \to A \iff \mathcal{H} \models_{\mathcal{K}} \Pi \to A.$$

As usual, $\Rightarrow$ stands for strong soundness (which usually holds for obvious reasons) and $\Leftarrow$ is strong completeness.

Strong completeness for the *product-free fragments* of $\mathbf{L}$ and $\mathbf{L}^\varepsilon$ was proved by Buszkowski [3] via a canonical model construction. Let $\mathbf{L}(\backslash, /)$ denote the calculus obtained from $\mathbf{L}$ by removing product from the set of operations and the corresponding rules from the set of rules; similarly for $\mathbf{L}^\varepsilon(\backslash, /)$.

Theorem 2 ([3]) *The calculi* $\mathbf{L}(\backslash, /)$ *and* $\mathbf{L}^\varepsilon(\backslash, /)$ *are strongly complete w.r.t., resp., L-models and Lε-models.*

Buszkowski's proof is very simple and elegant, so we present it below in Sect. 4 (even in a stronger form, with an extended set of operations). In contrast, Pentus' arguments for Theorem 1 (both for $\mathbf{L}$ and for $\mathbf{L}^\varepsilon$) are extremely complicated. This is connected to the fact that, in the presence of product, only weak completeness holds, while its strong version fails. Thus, any completeness proof should not be relativisable from 'pure' derivability to derivability from hypotheses, so it should use some intrinsic properties of the former.

The failure of strong completeness for $\mathbf{L}^\varepsilon$ was first noticed by Andréka and Mikulás [1, Prop. 4.2]:

Proposition 1 *Let $\mathcal{H}$ consist of two sequents: $p \to p \cdot p$ and $q \to p$. Then $\mathcal{H} \vDash_{\mathrm{L}\varepsilon\text{-models}} q \to p \cdot q$, but $\mathcal{H} \nvdash_{\mathbf{L}^\varepsilon} q \to p \cdot q$.*

Proof. For the first statement, consider an Lε-model $\mathcal{M}$ which satisfies $\mathcal{H}$. Since $\mathcal{M} \vdash p \to p \cdot p$, the language $\mathsf{w}(p)$ should either be empty or include the empty word. (Otherwise, we take the shortest word from $\mathsf{w}(p)$, of length $k > 0$, and see that it cannot belong to $\mathsf{w}(p \cdot p)$, where the shortest word has length $2k > k$.) Also, we have $\mathsf{w}(q) \subseteq \mathsf{w}(p)$.

If $\mathsf{w}(p) = \varnothing$, then $\mathsf{w}(q)$ is also empty, so $\mathsf{w}(q) \subseteq \mathsf{w}(p \cdot q)$. If $\varepsilon \in \mathsf{w}(p)$, then for any $u \in \mathsf{w}(q)$ we have $u = \varepsilon u \in \mathsf{w}(p \cdot q)$.

In order to prove that $q \to p \cdot q$ is not derivable from $\mathcal{H}$ in $\mathbf{L}^\varepsilon$, we use a syntactic argument.[2] The two extra axioms $p \to p \cdot p$ and $q \to p$, on top of $\mathbf{L}^\varepsilon$, can be equivalently replaced by the following two rules:

$$\frac{\Gamma, p, p, \Delta \to C}{\Gamma, p, \Delta \to C} \qquad \frac{\Gamma, p, \Delta \to C}{\Gamma, q, \Delta \to C}$$

(notice that here p and q are concrete variables, not meta-variables). Moreover, this new calculus enjoys cut elimination. Indeed, the only possibly problematic case in the cut elimination procedure is the one where the right premise is derived by one of the new rules, and the formula being cut is the "active" p or q. However, in this case the left premise of the cut, which is of the form $\Pi \to p$ or $\Pi \to q$, is either derived by a rule operating in Π, or is an axiom. In the first case, one propagates Cut to the left; in the second case, Cut disappears.

Now it is easy to see that $q \to p \cdot q$ has no cut-free derivation in the new calculus (and therefore it is not derivable from $\mathcal{H}$ in $\mathbf{L}^\varepsilon$). Indeed, one cannot immediately apply $\cdot R$, so the new rules should be applied. This can give sequents of the form $p, \ldots, p \to p \cdot q$. At some point, one has to apply $\cdot R$, whose left premise will be of the form $p, \ldots, p \to q$. This is obviously not derivable. □

The same counter-example works for **L** and the class of L-models. However, for this case Buszkowski [5, 8] gives a simpler example: $p \to p \cdot p \vDash_{\text{L-models}} p \to q$, but $p \to p \cdot p \nvdash_{\mathbf{L}} p \to q$ (in [5], q is replaced with $p \,/\, p$). Indeed, in the absence of ε validity of $p \to p \cdot p$ in a model means $\mathsf{w}(p) = \varnothing$, whence $\mathsf{w}(p) \subseteq \mathsf{w}(q)$, whatever $\mathsf{w}(q)$ is. The non-derivability can be established in the same syntactic way as above (while Buszkowski uses a semantic argument, using Lε-models).

The issue with product is in fact even deeper. Namely, the entailment relation on the class of Lε-models (or L-models), in the presence of product, is *not compact,* which makes it impossible to formulate a strongly sound and complete finitary calculus. Recall that compactness of an entailment relation means that entailment from an infinite set of hypotheses yields entailment from its sufficiently large finite subset. For derivability in a finitary calculus, compactness holds for obvious reasons, so a non-compact entailment relation cannot be axiomatised by such a calculus.

[2] Andréka and Mikulás prove it semantically, using another class of models—models on algebras of binary relations.

Proposition 2 *No finitary calculus, in a language including the product operation, can be strongly sound and complete w.r.t. the class of all L-models, and the same for the class of all Lε-models.*

Proof. We start with the case of L-models.

Let $\mathcal{L}$ be a calculus, which is strongly sound w.r.t. L-models. Let $\mathcal{H}$ be the following infinite set of sequents $\{q \to p^n \mid n \geq 1\}$ (recall that $p^n = p \cdot \ldots \cdot p$, n times). Then we claim that $\mathcal{H} \vDash_{\text{L-models}} q \to r$, but $\mathcal{H} \nvdash_{\mathcal{L}} q \to r$.

The first statement follows from the fact that in any L-model validating $\mathcal{H}$ the language $\mathsf{w}(q)$ is empty. Indeed, since there is no empty word, the length of each word in $\mathsf{w}(p^n)$ is at least n. Hence, there is no word belonging to $\mathsf{w}(p^n)$ for all n, and $\mathsf{w}(q)$ should be a subset of each $\mathsf{w}(p^n)$. Therefore, $\mathsf{w}(q) \subseteq \mathsf{w}(r)$ for any r.

For the second statement, we use a standard compactness argument. If $\mathcal{H} \vdash_{\mathcal{L}} q \to r$, then there exists a finite $\mathcal{H}_0 \subset \mathcal{H}$ with the same property. We may assume that $\mathcal{H}_0 = \{q \to p^n \mid 1 \leq n \leq m\}$ for some m. Now we construct a model $\mathcal{M}_m$ which validates $\mathcal{H}_0$ and falsifies $q \to r$. Let $\Sigma = \{a\}$ and let $\mathsf{w}(p) = \{a^n \mid n \geq 1\}$, $\mathsf{w}(q) = \{a^n \mid n \geq m\}$, and $\mathsf{w}(r) = \varnothing$. Obviously, $\mathcal{M}_m \nvDash q \to r$. On the other hand, $\mathsf{w}(p^k) = \{a^n \mid n \geq k\}$, whence $\mathcal{M}_m$ validates all sequents from $\mathcal{H}_0$. Since $\mathcal{L}$ is strongly sound w.r.t. L-models, we get $\mathcal{H}_0 \nvdash_{\mathcal{L}} q \to r$. Contradiction.

For the case of Lε-models, we take the same set $\mathcal{H} = \{q \to p^n \mid n \geq 1\}$ and $q \to p \cdot q$ as the goal sequent. Now if $\mathcal{M}$ validates $\mathcal{H}$, then there are two possibilities: either $\varepsilon \in \mathsf{w}(p)$ or $\mathsf{w}(q) = \varnothing$. In both cases, we have $\mathsf{w}(q) \subseteq \mathsf{w}(p \cdot q)$.

Now, again, if $\mathcal{H} \vdash_{\mathcal{L}} q \to p \cdot q$, we may assume that $\mathcal{H}_0 \vdash_{\mathcal{L}} q \to p \cdot q$, where $\mathcal{H}_0 = \{q \to p^n \mid 1 \leq n \leq m\}$ for some fixed m. The countermodel $\mathcal{M}_m$ is now constructed as follows. Let $\Sigma = \{a, b\}$ and let $\mathsf{w}(p) = \Sigma^+$ and $\mathsf{w}(q) = \{ab^k \mid k \geq m\}$. Since $\mathsf{w}(p^n) = \{u \in \Sigma^+ \mid |u| \geq n\}$, this model validates $\mathcal{H}_0$. On the other hand, the word ab^m belongs to $\mathsf{w}(q)$, but not to $\mathsf{w}(p \cdot q)$. Indeed, since $\varepsilon \notin \mathsf{w}(p)$, the $\mathsf{w}(q)$-part of this word is of the form b^k, but any word in $\mathsf{w}(q)$ should begin with an a. Hence, by strong soundness, $\mathcal{H}_0 \nvdash_{\mathcal{L}} q \to p \cdot q$. □

3 Extensions of the Lambek Calculus

Formal language semantics of the Lambek calculus motivates its extensions by means of other natural operations on formal languages. It is usually easy to write down the corresponding rules of inference, which guarantee strong soundness. However, as we shall see in the subsections below, completeness, even in the weak sense, could quickly break down.

3.1 Intersection and Union

The first most natural pair of operations are set-theoretical *intersection* and *union*. Following the tradition of Girard's linear logic [12], these two operations are also called *additive* conjunction and disjunction respectively (the product being *multiplicative* conjunction). The notion of L-models (Lε-models) is augmented with the new operations naturally:

$$\mathsf{w}(A \cap B) = \mathsf{w}(A) \cap \mathsf{w}(B); \qquad \mathsf{w}(A \cup B) = \mathsf{w}(A) \cup \mathsf{w}(B).$$

The rules of inference (the same for **L** and for $\mathbf{L}^\varepsilon$) are as follows. The resulting systems are two variants of the *multiplicative-additive Lambek calculus* (see [14]); we denote them by **MALC** and $\mathbf{MALC}^\varepsilon$.

$$\frac{\Gamma, A, \Delta \to C}{\Gamma, A \cap B, \Delta \to C}\ \cap L_1 \qquad \frac{\Gamma, B, \Delta \to C}{\Gamma, A \cap B, \Delta \to C}\ \cap L_2 \qquad \frac{\Pi \to A \quad \Pi \to B}{\Pi \to A \cap B}\ \cap R$$

$$\frac{\Gamma, A, \Delta \to C \quad \Gamma, B, \Delta \to C}{\Gamma, A \cup B, \Delta \to C}\ \cup L \qquad \frac{\Pi \to A}{\Pi \to A \cup B}\ \cup R_1 \qquad \frac{\Pi \to B}{\Pi \to A \cup B}\ \cup R_2$$

The principal issue with completeness, in the presence of $\cap$ and $\cup$, comes from the *distributivity law:*

$$(A \cup C) \cap (B \cup C) \to (A \cap B) \cup C.$$

This sequent is obviously true in all L-models (resp., Lε-models). However, it is not derivable in **MALC** or $\mathbf{MALC}^\varepsilon$, which can be easily established by exhaustive proof search. (In fact, this is a general issue for substructural logics lacking contraction and weakening, see [30].)

A natural question arises whether adding the distributivity law would give a complete system. The distributivity law could be added as an additional axiom scheme, and in this case the *Cut* rule should be included into the system. There also exists a cut-free sequent calculus for $\mathbf{MALC}^\varepsilon$ with distributivity, due to Kozak [18]. In Kozak's system, antecedents of sequents have a richer structure, allowing two metaconnectives: the usual comma for product, and another one—say, semicolon—for additive conjunction, and they may be interleaved. For semicolon, structural rules of contraction and weakening are allowed. Completeness w.r.t. L-models (Lε-models), even in the weak sense, remains an open question.

Interestingly enough, however, intersection and union behave differently w.r.t. completeness. On one hand, the issue with distributivity remains in the fragment with the following connectives: \, /, $\cup$. In other words, only $\cup$ is sufficient to formulate a sequent which is not derivable in $\mathbf{MALC}^\varepsilon$, but becomes derivable in the presence of distributivity (and in particular, is true in all Lε-models). Such a sequent is presented in [16]:

Proposition 3 *The following sequent is not derivable in* $\mathbf{MALC}^\varepsilon$ *(and thus also in* **MALC***), but follows from the distributivity law, and is therefore true in all L-models and in all Lε-models:*

$$((p \mathbin{/} q) \cup s) \mathbin{/} ((p \mathbin{/} q) \cup (p \mathbin{/} r) \cup s), (p \mathbin{/} q) \cup s, ((p \mathbin{/} q) \cup s) \setminus ((p \mathbin{/} r) \cup s) \\ \to (p \mathbin{/} (q \cup r)) \cup s.$$

Looking weird at the first glance, this sequent has a natural origin. One starts with the distributivity law for $A = p \mathbin{/} q$, $B = p \mathbin{/} r$ and $C = s$. On the right, $(A \cap B) \cup C$, where $A \cap B$ gets replaced by equivalent $p \mathbin{/} (q \cup r)$, getting rid of $\cap$. On the left, instead of $(A \cup C) \cap (B \cup C)$, we use the diamond construction (see [32]): since $A \cup C \to A \cup B \cup C$ and $B \cup C \to A \cup B \cup C$, for the sequence $\Pi = (A \cup C) \mathbin{/} (A \cup B \cup C), A \cup C, (A \cup C) \setminus (B \cup C)$ we have both $\Pi \to A \cup C$ and $\Pi \to B \cup C$. Thus, Π can act as a substitute for $(A \cup C) \cap (B \cup C)$, so the sequent in question follows from the distributivity law. The fact that it is not derivable in $\mathbf{MALC}^{\varepsilon}$ is established by exhaustive proof search (see Kanovich et al. [16]).

On the other hand, the fragment of **MALC** (resp., $\mathbf{MALC}^{\varepsilon}$) with only $\setminus$, $/$, and $\cap$ is strongly complete w.r.t. L-models (resp., Lε-models), which is established by Buszkowski's canonical model construction (Sect. 4 below). For the fragment also including the product (i.e., **MALC** or $\mathbf{MALC}^{\varepsilon}$ without $\cup$), weak completeness remains an open problem. In the presence of product, strong completeness fails (see above), so in order to prove weak completeness one has to extend Pentus' arguments to the case with intersection, which is probably technically hard.

3.2 *Iteration (Kleene Star)*

Another natural operation on formal languages which can be added to the Lambek calculus is *iteration,* also called *Kleene star*. The iteration of a formal language is defined as follows:

$$M^* = \{u_1 \ldots u_n \mid n \geq 0;\, u_1, \ldots, u_n \in M\} = \bigcup_{n=0}^{\infty} M^n.$$

The rules for Kleene star, extending $\mathbf{MALC}^{\varepsilon}$, are as follows:

$$\frac{\left(\Gamma, A^n, \Delta \to C\right)_{n=0}^{\infty}}{\Gamma, A^*, \Delta \to C}\ {}^*L \qquad \frac{\Pi_1 \to A \quad \ldots \quad \Pi_n \to A}{\Pi_1, \ldots, \Pi_n \to A^*}\ {}^*R,\ n \geq 0$$

(Here A^n means $A, \ldots, A$, with A repeated n times.)

The ${}^*L$ rule is an ω*-rule:* it has an infinite sequence of premises, ordered by natural numbers (i.e., of order type ω). Proofs where applications of ω-rules are allowed can be infinite trees, but are still required to be well-founded: no infinite paths going upwards from the root are allowed. Alternatively, one may define the set of theorems (derivable sequents) as the smallest (by inclusion) set which includes all axiom instances and is closed under inference rules.

The aforementioned extension of $\mathbf{MALC}^{\varepsilon}$ with Kleene star was introduced by Buszkowski and Palka [7, 9, 31]. This extension is denoted by $\mathbf{ACT}_{\omega}$ and called *infinitary action logic*. The name comes from the term "action lattices," which is an alternative name for residuated Kleene lattices, see [19, 35]. Using a transfinite version of the standard argument, Palka proved cut elimination for $\mathbf{ACT}_{\omega}$ [31, Th. 3.1].

The derivability problem for $\mathbf{ACT}_{\omega}$ is Π_1^0-complete [7, 31]. In particular, the set of theorems of $\mathbf{ACT}_{\omega}$ is not recursively enumerable, and therefore it cannot be axiomatised by a finitary calculus. The usage of some sort of infinitary proof mechanisms is inevitable: those could be the ω-rule or, for example, non-well-founded proofs [11].

The axiomatisation for Kleene star essentially requires empty antecedents to be allowed (in particular, ${}^*R$ for $n = 0$ is just $\to A^*$); respectively, each language M^* necessarily includes the empty word. In L-models without ε, Kleene star is replaced by a closely related operation called *positive iteration* ("Kleene plus"):

$$M^+ = \{u_1 \dots u_n \mid n \geq 1; u_1, \dots, u_n \in M\} = \bigcup_{n=1}^{\infty} M^n.$$

The rules for positive iteration, being added to **MALC**, are as follows:

$$\frac{\left(\Gamma, A^n, \Delta \to C\right)_{n=1}^{\infty}}{\Gamma, A^+, \Delta \to C}\ {}^+L \qquad \frac{\Pi_1 \to A \quad \dots \quad \Pi_n \to A}{\Pi_1, \dots, \Pi_n \to A^+}\ {}^+R,\ n \geq 1$$

(the left rule is again an ω-rule). The extension of **MALC** with the aforementioned rule is denoted by $\mathbf{ACT}_{\omega}^{+}$ (positive infinitary action logic).

In the absence of Lambek's non-emptiness restriction, i.e., in $\mathbf{ACT}_{\omega}$, there is no need to add positive iteration as a separate operation, as it is easily expressible using Kleene star: $A^+ = A \cdot A^*$.

As one may notice from the definition, Kleene star is a special sort of (infinite) union. This leads to incompleteness w.r.t. Lε-models (or L-models for positive iteration), for fragments including product, division, intersection, and iteration. This issue is again connected to distributivity. The examples presented below were introduced in [21, 25].

Proposition 4 *The sequent*

$$(s \,/\, (r \,/\, r)) \cap (s \,/\, (p \cdot p^* \cap q \cdot q^*) \to s \,/\, (p^* \cap q^*)$$

is true in all Lε-models, but is not derivable in $\mathbf{ACT}_{\omega}$.
The sequent

$$(s \,/\, (p \cap q)) \cap (s \,/\, (p \cap q \cdot q^+)) \cap (s \,/\, (p \cdot p^+ \cap q)) \cap (s \,/\, (p \cdot p^+ \cap q \cdot q^+))$$
$$\to s \,/\, (p^+ \cap q^+)$$

is true in all L-models, but is not derivable in $\mathbf{ACT}_\omega^+$.[3]

For the first sequent, we use distributivity and derivable sequents of the form $A^* \to (r \,/\, r) \cup A \cdot A^*$, where $r \,/\, r$ is a substitute for the unit constant representing $\{\varepsilon\}$ (see below). The second sequent here uses the fact that A^+ is equivalent to $A \cup A \cdot A^+$, and $(s \,/\, A_1) \cap \ldots \cap (s \,/\, A_4)$ is equivalent to $s \,/\, (A_1 \cup A_2 \cup A_3 \cup A_4)$. After such replacements, the sequents becomes an instance of the distributivity law under $s \,/ \ldots$

Thus, both sequents follow from the distributivity laws, and therefore they are true on algebras of formal languages. The fact that they are not derivable in $\mathbf{ACT}_\omega$ or, respectively, $\mathbf{ACT}_\omega^+$ is established by exhaustive proof search (which happens to be possible despite the calculi are in general undecidable).

3.3 Iterative Divisions

In order to overcome this incompleteness issue, the following restricted versions of iteration were introduced [25]. In the restricted systems, iteration is allowed to be used only in denominators of divisions. In other words, Kleene star is replaced by two operations: $A^* \setminus B$ and $B \,/\, A^*$ (each one is *one binary operation* with operands A and B); for positive iteration, the new operations are $A^+ \setminus B$ and $B \,/\, A^+$. These operations are called *iterative divisions* and were introduced in the context of $\mathbf{ACT}_\omega^+$ in [25]. Independently, they were defined in a non-associative setting by Sedlár [36].

The rules for iterative divisions are as follows:

$$\frac{\Pi_1 \to A \quad \ldots \quad \Pi_n \to A \quad \Gamma, B, \Delta \to C}{\Gamma, \Pi_1, \ldots, \Pi_n, A^* \setminus B, \Delta \to C}\ {}^*\!\setminus L, n \geq 0 \qquad \frac{\left(A^n, \Pi \to B\right)_{n=0}^{\infty}}{\Pi \to A^* \setminus B}\ {}^*\!\setminus R$$

$$\frac{\Pi_1 \to A \quad \ldots \quad \Pi_n \to A \quad \Gamma, B, \Delta \to C}{\Gamma, B \,/\, A^*, \Pi_1, \ldots, \Pi_n, \Delta \to C}\ /^* L, n \geq 0 \qquad \frac{\left(\Pi, A^n \to B\right)_{n=0}^{\infty}}{\Pi \to B \,/\, A^*}\ /^* R$$

(Here the right rules are ω-rules, and the left ones are finitary.)

The versions of the aforementioned rules for the positive setting are as follows:

$$\frac{\Pi_1 \to A \quad \ldots \quad \Pi_n \to A \quad \Gamma, B, \Delta \to C}{\Gamma, \Pi_1, \ldots, \Pi_n, A^+ \setminus B, \Delta \to C}\ {}^+\!\setminus L, n \geq 1 \qquad \frac{\left(A^n, \Pi \to B\right)_{n=1}^{\infty}}{\Pi \to A^+ \setminus B}\ {}^+\!\setminus R$$

$$\frac{\Pi_1 \to A \quad \ldots \quad \Pi_n \to A \quad \Gamma, B, \Delta \to C}{\Gamma, B \,/\, A^+, \Pi_1, \ldots, \Pi_n, \Delta \to C}\ /^+ L, n \geq 1 \qquad \frac{\left(\Pi, A^n \to B\right)_{n=1}^{\infty}}{\Pi \to B \,/\, A^+}\ /^+ R$$

[3] Here we suppose that iteration has higher priority than product, and they both have higher priority over other operations.

Obviously, in the presence of the full-power iteration operations (Kleene star or positive iteration, respectively), these new rules just combine the rules for iteration and division (i.e., they are derivable in $\mathbf{ACT}_\omega$ or $\mathbf{ACT}^+_\omega$). If we consider, however, fragments of our calculi where only iterative divisions, but not other uses of iteration, are allowed, these rules provide a self-contained axiomatisation.

In systems which include both iteration and iterative divisions, with their own sets of rules, *Cut* is eliminable [25]. Therefore, for derivability without hypotheses the given rules define *conservative fragments* of $\mathbf{ACT}_\omega$ and $\mathbf{ACT}^+_\omega$: for sequents in the restricted language, derivability using only the new rules is equivalent to derivability in the full system. Moreover, conservativity also holds for deriving from hypotheses (*strong conservativity*), which is proved by a cut normalisation argument [23].

Unlike Kleene star or positive iteration, which are infinite unions, iterative divisions can be represented as infinite *intersections:*

$$M^* \setminus N = \bigcap_{n=0}^{\infty} (M^n \setminus N) \qquad N \,/\, M^* = \bigcap_{n=0}^{\infty} (N \,/\, M^n)$$

$$M^+ \setminus N = \bigcap_{n=1}^{\infty} (M^n \setminus N) \qquad N \,/\, M^+ = \bigcap_{n=1}^{\infty} (N \,/\, M^n)$$

As noticed above, in contrast with union, intersection does not create obstacles to completeness—and indeed, the fragments with iterative divisions are strongly complete w.r.t. L-models (Lε-models). This is proved by the same Buszkowski's canonical model construction, see Sect. 4 below.

3.4 *Constants* **0** *and* **1**

Finally, we extend the Lambek calculus with two constants: *zero* and *unit*. The zero constant can be added both to $\mathbf{L}$ and $\mathbf{L}^\varepsilon$ (and their aforementioned extensions); the natural interpretation of zero is the empty language: $\mathsf{w}(\mathbf{0}) = \varnothing$. As for the unit, its interpretation is $\mathsf{w}(\mathbf{1}) = \{\varepsilon\}$, which is available only in Lε-models. Hence, we add $\mathbf{1}$ only to systems without Lambek's non-emptiness restriction, i.e., $\mathbf{L}^\varepsilon$ and its extensions.

The only rule for $\mathbf{0}$ is the following left rule (axiom); no right rule is added.

$$\frac{}{\Gamma, \mathbf{0}, \Delta \to C}\ \mathbf{0}L$$

For $\mathbf{1}$, the rules are as follows:

$$\frac{\Gamma, \Delta \to C}{\Gamma, \mathbf{1}, \Delta \to C}\ 1L \qquad \frac{}{\to \mathbf{1}}\ 1R$$

In the presence of the unit constant, one can alternatively define $\mathbf{ACT}_\omega$ in terms of positive iteration, and then define Kleene star as $A^* = \mathbf{1} \cup A^+$.

The conditions $\mathsf{w}(\mathbf{0}) = \varnothing$ and $\mathsf{w}(\mathbf{1}) = \{\varepsilon\}$ define the *standard interpretation* of constants $\mathbf{0}$ and $\mathbf{1}$ on algebras of formal languages. Unfortunately, both constants ruin completeness, even in the weak sense. In order to show this, consider formulae of the form $\mathbf{0} \mathbin{/} p$ and $\mathbf{1} \mathbin{/} p$. As one can easily see $\mathsf{w}(\mathbf{0} \mathbin{/} p)$ is either $\varnothing$ or Σ^* (for L-models, Σ^+), and $\mathsf{w}(\mathbf{1} \mathbin{/} p)$ can have three possible values: $\varnothing$, $\{\varepsilon\}$, Σ^*. Thus, on formulae of the aforementioned form semantically we have, respectively, a two-valued or a three-valued logic (see [22] for details), which validates more sequents than the Lambek calculus with constants. As an example, consider the commutativity principles:

$$\mathbf{0} \mathbin{/} p, \mathbf{0} \mathbin{/} q \to (\mathbf{0} \mathbin{/} q) \cdot (\mathbf{0} \mathbin{/} p) \qquad \mathbf{1} \mathbin{/} p, \mathbf{1} \mathbin{/} q \to (\mathbf{1} \mathbin{/} q) \cdot (\mathbf{1} \mathbin{/} p)$$

These principles are semantically valid, but not derivable (which is established by exhaustive proof search).

As noticed below in Sect. 4, completeness, in the strong form, can be recovered by considering non-standard interpretations of $\mathbf{0}$ and $\mathbf{1}$. For the standard interpretation of $\mathbf{1}$, however, as we show in Sect. 5, the incompleness issue is in fact very deep. In particular, in the presence of $\mathbf{1}$, $\cap$ and $\backslash$, the set of sequents generally true in Lε-models is undecidable (without $\mathbf{1}$ it coincides with the set of theorems of the corresponding fragment of $\mathbf{MALC}^\varepsilon$, which is decidable). Moreover, if one adds also Kleene star, we conjecture Π^1_1-completeness.

4 Buszkowski's Canonical Models

In this section, we extend Buszkowski's strong completeness result (Theorem 2) to the language with intersection and iterative divisions, following [25].

Theorem 3 *The fragment of* $\mathbf{ACT}_\omega$ *with the following connectives:* $\backslash$, $/$, ${}^*\backslash$, $/^*$, *and* $\cap$ *is strongly complete w.r.t. L*ε*-models. The fragment of* $\mathbf{ACT}^+_\omega$ *with the following connectives:* $\backslash$, $/$, ${}^+\backslash$, $/^+$, *and* $\cap$ *is strongly complete w.r.t. L-models.*

Here, in the absence of iteration, we regard iterative divisions (${}^*\backslash$, $/^*$; ${}^+\backslash$, $/^+$) as independent connectives, governed by inference rules given in the previous section.

Proof. The argument is the same with and without Lambek's restriction. To be definite, let us consider the case of $\mathbf{ACT}_\omega$.

Let $\mathcal{H}$ be a set of sequents (finite or infinite) which will be used as hypotheses. We shall construct a *universal model* $\mathcal{M}_\mathcal{H}$, which will be an Lε-model which validates *exactly* the sequents which are derivable from $\mathcal{H}$ in $\mathbf{ACT}_\omega$. Thus, for a sequent $\Pi \to A$ such that $\mathcal{H} \nvdash_{\mathbf{ACT}_\omega} \Pi \to A$, our model $\mathcal{M}_\mathcal{H}$ will serve as the desired countermodel establishing $\mathcal{H} \nvDash \Pi \to A$. Indeed, it validates all sequents from $\mathcal{H}$ and falsifies $\Pi \to A$.

As the alphabet Σ for $\mathcal{M}$, we take the set of all formulae in the restricted language, i.e., those built from variables using $\backslash$, $/$, ${}^*\backslash$, $/^*$, and $\cap$. The valuation function is defined in the following natural way:

$$\mathsf{w}(A) = \{\Gamma \mid \mathcal{H} \vdash_{\mathbf{ACT}_\omega} \Gamma \to A\}.$$

Let us show that $\mathcal{M} = (\Sigma, \mathsf{w})$ is indeed an Lε-model, that is, that w correctly commutes with the operations. (In what follows, we write just $\vdash$ instead of $\vdash_{\mathbf{ACT}_\omega}$.)

Division. Let $\Gamma \in \mathsf{w}(A \backslash B)$. In order to show that $\Gamma \in \mathsf{w}(A) \backslash \mathsf{w}(B)$, we take an arbitrary $\Delta \in \mathsf{w}(A)$ and show that $\Delta\Gamma \in \mathsf{w}(B)$. Indeed, we have $\mathcal{H} \vdash \Delta \to A$ and $\mathcal{H} \vdash \Gamma \to A \backslash B$, and now the desired $\Delta, \Gamma \to B$ is derived as follows:

$$\dfrac{\Gamma \to A \backslash B \qquad \dfrac{\Delta \to A \quad B \to B}{\Delta, A \backslash B \to B}\,\backslash L}{\Delta, \Gamma \to B}\,Cut$$

For the opposite direction, let $\Gamma \in \mathsf{w}(A) \backslash \mathsf{w}(B)$ and take $A \in \mathsf{w}(A)$ (since $A \to A$ is an axiom). Then $A\Gamma \in \mathsf{w}(B)$, whence $\mathcal{H} \vdash A, \Gamma \to B$. By $\backslash R$, we derive $\Gamma \to A \backslash B$, therefore $\Gamma \in \mathsf{w}(A \backslash B)$. The $/$ case is symmetric.

Intersection. Here we have

$$\mathsf{w}(A \cap B) = \{\Gamma \mid \mathcal{H} \vdash \Gamma \to A \cap B\} = \{\Gamma \mid \mathcal{H} \vdash \Gamma \to A \text{ and } \mathcal{H} \vdash \Gamma \to B\} = \mathsf{w}(A) \cap \mathsf{w}(B).$$

The second equality is due to the $\cap R$ rule (for $\supseteq$) and by Cut with $A \cap B \to A$ and $A \cap B \to B$ (for $\subseteq$).

Iterative division. In fact, as noticed above, $A^* \backslash B$ can be represented as an infinite intersection $\bigcap_{n=0}^{\infty}(A^n \backslash B)$, and each $A^n \backslash B$ is a series of divisions: $A \backslash (\ldots (A \backslash B) \ldots)$. Thus, there is no surprise that $\mathsf{w}(A^* \backslash B)$ is $\mathsf{w}(A)^* \backslash \mathsf{w}(B)$.

Let us give an accurate proof. Take $\Gamma \in \mathsf{w}(A^* \backslash B)$ and an arbitrary $\Delta \in \mathsf{w}(A)^*$. The latter means that $\Delta = \Delta_1, \ldots, \Delta_n$ for some $n \geq 0$, and each Δ_i belongs to $\mathsf{w}(A)$. Now we show that $\Delta\Gamma \in \mathsf{w}(B)$ by the following derivation:

$$\dfrac{\Gamma \to A^* \backslash B \qquad \dfrac{\Delta_1 \to A \quad \ldots \quad \Delta_n \to A \quad B \to B}{\Delta_1, \ldots, \Delta_n, A^* \backslash B \to B}\,{}^*\backslash L}{\Delta_1, \ldots, \Delta_n, \Gamma \to B}\,Cut$$

(In particular, for $n = 0$ there is an instance of ${}^*\backslash L$ which derives $A^* \backslash B \to B$ from $B \to B$.)

For the opposite direction, we notice that $\mathsf{w}(A)^*$ contains A^n for each $n \geq 0$. Therefore, if $\Gamma \in \mathsf{w}(A)^* \backslash \mathsf{w}(B)$, then $A^n\Gamma \in \mathsf{w}(B)$ for each $n \geq 0$, i.e., each sequent

$A^n, \Gamma \to B$ is derivable from $\mathcal{H}$. By the ω-rule $^*\backslash R$, we derive $\Gamma \to A^* \setminus B$, whence $\Gamma \in \mathsf{w}(A^* \setminus B)$.

The $/^*$ case is symmetric.

The changes for the case of $\mathbf{ACT}^+_\omega$ are minor. Due to Lambek's non-emptiness restriction, each $\mathsf{w}(A)$ does not include the empty word (empty sequence of formulae). In particular, this means that the usage of $\backslash R$ in the division case is correct. For iterative divisions, $^*\backslash$ is replaced by $^+\backslash$, and $n \geq 0$ is replaced by $n \geq 1$. □

The models $\mathcal{M}_\mathcal{H}$ can be naturally called *Buszkowski's canonical models.* As noticed above, this canonical model construction does not extend to product, union and iteration, neither does it to the zero and unit constants. Indeed, otherwise the corresponding extensions of the Lambek calculus would have been strongly complete w.r.t. L-models (Lε-models), which is not the case (see above).

In fact, the operations for which Buszkowski's canonical model works, namely, $\backslash$, $/$, $^*\backslash$, $/^*$ (resp., $^+\backslash$ and $/^+$), and $\cap$, are exactly those for which the right logical rules are *invertible* (in Andreoli's focusing [2] terminology, *asynchronous*). Where this is not the case, the argument fails. For example, for product we would have $\mathsf{w}(A) \cdot \mathsf{w}(B) \subseteq \mathsf{w}(A \cdot B)$ due to the $\cdot R$ rule, but in general $\mathsf{w}(A \cdot B) \not\subseteq \mathsf{w}(A) \cdot \mathsf{w}(B)$. Indeed, $p\,(p \setminus (q \cdot r))$ belongs to $\mathsf{w}(q \cdot r)$, but cannot be represented as a concatenation of words from $\mathsf{w}(q)$ and $\mathsf{w}(r)$ (recall that $(p \setminus (q \cdot r))$ is regarded as one letter). The reason is that in a derivation of $p, p \setminus (q \cdot r) \to q \cdot r$ the $\cdot R$ rule cannot be applied immediately; first one has to use $\backslash L$.

Constants $\mathbf{0}$ and $\mathbf{1}$ also ruin Buszkowski's canonical model construction, for the same reason: a sequent $\Pi \to \mathbf{1}$ can be derivable even if it is not the $\to \mathbf{1}$ axiom, and $\Pi \to \mathbf{0}$ can be derivable, despite there is no right rule for $\mathbf{0}$. Thus, $\mathsf{w}(\mathbf{1}) \neq \{\varepsilon\}$ and $\mathsf{w}(\mathbf{0}) \neq \varnothing$. A modified completeness result, however, may be extracted from the canonical model construction. Namely, we consider models with *non-standard* interpretations of $\mathbf{0}$ and $\mathbf{1}$, where $\mathsf{w}(\mathbf{0})$ should be just the smallest (by inclusion) set among sets of the form $\mathsf{w}(A)$, and $\mathsf{w}(\mathbf{1})$ should be the neutral by product, again among languages of the form $\mathsf{w}(A)$. For such generalised interpretations, we obtain strong completeness.

The alphabets used in canonical models $\mathcal{M}_\mathcal{H}$ are infinite ones. They could be reduced to finite ones in the case of a finite $\mathcal{H}$ (by taking only subformulae of $\mathcal{H}$ and the goal sequent $\Pi \to A$). Moreover, a stronger result holds: as shown by Pentus [33], a countable alphabet can be uniformly reduced to a two-letter one. (This is the smallest possible alphabet, since product of languages over a one-letter alphabet is commutative, so completeness in this case would definitely fail.)

Theorem 4 *The fragment of* $\mathbf{ACT}_\omega$ *with the following connectives:* $\backslash$, $/$, $^*\backslash$, $/^*$, *and* $\cap$ *is strongly complete w.r.t. Lε-models over alphabet* $\Sigma_2 = \{b, c\}$. *The fragment of* $\mathbf{ACT}^+_\omega$ *with the following connectives:* $\backslash$, $/$, $^+\backslash$, $/^+$, *and* $\cap$ *is strongly complete w.r.t. L-models over alphabet* $\Sigma_2 = \{b, c\}$.

Proof. The same argument is used for both cases, with and without Lambek's restriction.

Let $\Sigma = \{a_1, a_2, \ldots\}$ be the alphabet of the models $\mathcal{M}_{\mathcal{H}}$ from the proof of Theorem 3, i.e., the set of formulae with the given set of operations. Consider the homomorphism $h \colon \Sigma^* \to \Sigma_2^*$, defined on letters as follows:

$$h(a_i) = bc^i b$$

(and propagated to arbitrary words as usual: $h(\varepsilon) = \varepsilon$, $h(uv) = h(u)h(v)$). It is easy to see that h is injective.

Given the model $\mathcal{M}_{\mathcal{H}} = (\Sigma, \mathsf{w})$, for a given set of hypotheses $\mathcal{H}$, we construct a new model $\mathcal{M}'_{\mathcal{H}} = (\Sigma_2, \mathsf{w}')$, over the two-letter alphabet, by the following valuation function:

$$\mathsf{w}'(A) = h(\mathsf{w}(A)).$$

Being a language-theoretic homomorphism, h commutes with product and intersection, therefore, so does the new valuation function w'. Moreover, h also commutes with the infinite union operation, therefore $h(M^*) = (h(M))^*$ for any language M.

The more interesting case is division. We claim that $h(M \setminus N) = h(M) \setminus h(N)$, if $M \neq \varnothing$.[4] For the $\subseteq$ inclusion, take $u' \in h(M \setminus N)$ and an arbitrary $v' \in h(M)$, i.e., $u' = h(u)$, $v' = h(v)$, $u \in M \setminus N$, and $v \in M$. We have $vu \in N$ and therefore $v'u' = h(vu) \in h(N)$. Hence, $u' \in h(M) \setminus h(N)$. For the $\supseteq$ inclusion, let $u' \in h(M) \setminus h(N)$. Since $M \neq \varnothing$, we may take an element $v_0 \in M$, and we have $h(v_0)\, u' \in h(N)$. Therefore, $h(v_0)\, u'$ is a word of the form $bc^{i_1}bbc^{i_2}b \ldots bc^{n}b$, and therefore so is u'. In other words, $u' = h(u)$ for some $u \in \Sigma^*$. Now let us show that $u \in M \setminus N$, which immediately yields the desired $u' = h(u) \in h(M \setminus N)$. Indeed, for any $v \in M$ we have $h(v) \in h(M)$, and therefore $h(vu) = h(v)\, u' \in h(N)$. Since h is injective, this entails the needed $vu \in N$. The / case is symmetric.

Notice that $\mathsf{w}(A)$ is non-empty for any A, as it includes (at least) the one-letter word A. Therefore, for any A and B we have $\mathsf{w}'(A \setminus B) = h(\mathsf{w}(A \setminus B)) = h(\mathsf{w}(A) \setminus \mathsf{w}(B)) = h(\mathsf{w}(A)) \setminus h(\mathsf{w}(B)) = \mathsf{w}'(A) \setminus \mathsf{w}'(B)$, and similarly for $B \,/\, A$.

Thus, the new valuation function w' correctly defines a model, which we have called $\mathcal{M}'_{\mathcal{H}}$. Moreover, $\mathcal{M}'_{\mathcal{H}}$ validates exactly the same sequents as $\mathcal{M}_{\mathcal{H}}$ does; this is due to injectivity of h. In particular, if $\mathcal{H} \nvdash \Pi \to A$, then $\mathcal{M}'_{\mathcal{H}}$ serves as a countermodel to show that $\mathcal{H} \nvDash \Pi \to A$. This establishes strong completeness w.r.t. models over Σ_2. □

5 The Unit Constant

In this final section, we discuss incompleteness issues connected to the unit constant more deeply. Namely, we concentrate on formulae of the form $\mathbf{1} \cap A$ and show that they, in Lε-models, satisfy the structural rules (weakening, contraction, permutation),

[4] Notice that this constraint is crucial, since $\varnothing \setminus h(N)$ includes all words, including those which are not of the form $h(v)$ for some $v \in \Sigma^*$.

which are not granted by the Lambek calculus, as the latter is a substructural logical system. This fact not only leads to incompleteness, but also allows to prove an algorithmic undecidability result [16].

The key lemma is the following statement, which is a form of deduction theorem, in terms of entailment on the class of Lε-models and the "$\mathbf{1}\cap$" modifier for the hypotheses. The formulae are considered in the largest possible language, i.e., they may use all operations of $\mathbf{ACT}_\omega$.

Lemma 1 *Let A and B be two formulae. Then $\to A \vDash_{\mathrm{L}\varepsilon-\text{models}} \to B$ if and only if $\mathbf{1} \cap A \to B$ is true in all Lε-models.*

This lemma and the next one were stated (without proof) by Buszkowski [6], where they are used to show Σ_1^0-hardness of the equational theory of Lε-models in the presence of $\mathbf{1}$ (Theorem 5 below). Let us give accurate proofs.

Proof. For the "if" direction, let $\mathbf{1} \cap A \to B$ be true in all Lε-models and consider an Lε-model $\mathcal{M} = (\Sigma, \mathsf{w})$ such that $\mathcal{M} \vDash\ \to A$. The latter means $\varepsilon \in \mathsf{w}(A)$, and therefore $\varepsilon \in \mathsf{w}(\mathbf{1} \cap A)$. By validity of $\mathbf{1} \cap A \to B$ in $\mathcal{M}$, we have $\mathsf{w}(\mathbf{1} \cap A) \subseteq \mathsf{w}(B)$, whence $\varepsilon \in \mathsf{w}(B)$. This means that $\to B$ is true in $\mathcal{M}$.

For the "only if" direction, consider an arbitrary Lε-model $\mathcal{M} = (\Sigma, w)$. If $\varepsilon \notin \mathsf{w}(A)$, then $\mathsf{w}(\mathbf{1} \cap A) = \varnothing$, and therefore $\mathbf{1} \cap A \to B$ is trivially true in $\mathcal{M}$. Otherwise, $\mathcal{M}$ validates $\to A$ and, by $\to A \vDash_{\mathrm{L}\varepsilon-\text{models}} \to B$, $\mathcal{M}$ also validates $\to B$. Thus, we have $\mathsf{w}(\mathbf{1} \cap A) = \{\varepsilon\}$ and $\varepsilon \in \mathsf{w}(B)$, and $\mathcal{M}$ again validates $\mathbf{1} \cap A \to B$. □

Since the intersection operation acts as conjunction, this lemma can be extended to entailment from finite set of hypotheses.

Lemma 2 *Let $\mathcal{H} = \{\to A_1, \ldots, \to A_n\}$ be a set of hypotheses with empty antecedents and let B be a formula. Then $\mathcal{H} \vDash_{\mathrm{L}\varepsilon-\text{models}} \to B$ if and only if $\mathbf{1} \cap A_1 \cap \ldots \cap A_n \to B$ is true in all Lε-models.*

Proof. The sequent $\to A_1 \cap \ldots \cap A_n$ is true in an Lε-model $\mathcal{M}$ if and only if all sequents $\mathcal{H}$ are true in $\mathcal{M}$ simultaneously. Hence, we may replace $\mathcal{H}$ with $\{\to A_1 \cap \ldots \cap A_n\}$ and then apply Lemma 1. □

This lemma leads to an alternative proof of the undecidability result [16, Thm. 8] for the set of sequents, in the language of $\backslash$, $\cap$, and $\mathbf{1}$, which are generally true in Lε-models:

Theorem 5 *The set of sequents, in the language of $\backslash$, $\cap$, and $\mathbf{1}$, which are generally true in Lε-models, is undecidable, more precisely, Σ_1^0-hard.*

Proof. We build upon undecidability of derivability from finite sets of hypotheses in the one-division fragment of the Lambek calculus (i.e., for sequents built using only $\backslash$), proved by Buszkowski [4] (see also [15] for the version without Lambek's restriction). This problem is Σ_1^0-complete, as it encodes derivability in type-0 generative grammars. We may suppose that all sequents in questions have empty antecedents:

$F_1, \ldots, F_n \to G$ can be equivalently replaced by $\to F_n \setminus (\ldots (F_1 \setminus G) \ldots)$. Thus, a given set of hypotheses $\mathcal{H}$ is of the form $\mathcal{H} = \{ \to A_1, \ldots, \to A_n\}$.

Next, we use Lemma 2 in combination with Theorem 2 (strong completeness of the product-free fragment of $\mathbf{L}^{\varepsilon}$) in order to reduce derivability from hypotheses to general validity in the presence of $\mathbf{1}$ and $\cap$, over the class of Lε-models:

$$\mathcal{H} \vdash_{\mathbf{L}^{\varepsilon}} \to B \iff \mathcal{H} \models_{\mathrm{L}\varepsilon\text{-models}} \to B \iff \mathbf{1} \cap A_1 \cap \ldots \cap A_n \to B \text{ is generally true.}$$

Here the usage of strong completeness is correct, since $\mathcal{H}$ and B are built using only $\setminus$, and do not include the unit constant. □

Notice that this theorem establishes only a lower bound. No weakly complete calculus for the language of $\setminus, \cap, \mathbf{1}$ is known, which could have guaranteed enumerability of the logic in question. It is quite likely that its complexity is in fact higher than Σ_1^0.

The deeper idea behind this undecidability result is that $\mathbf{1} \cap A$, in Lε-models, actually behaves as the *exponential modality* of Girard's linear logic, $!A$ (see [12, 13, 28]). Using the exponential, one encodes derivability from hypotheses (via a modalised version of deduction theorem), and thus obtains undecidability results in the spirit of the work of Lincoln et al. [28].

Recall that the exponential is a modality (unary connective) ! governed by the following logical and structural rules:

$$\frac{\Gamma, A, \Delta \to C}{\Gamma, !A, \Delta \to C}\ !L \qquad \frac{!A_1, \ldots, !A_n \to B}{!A_1, \ldots, !A_n \to !B}\ !R$$

$$\frac{\Gamma, \Delta \to C}{\Gamma, !A, \Delta \to C}\ !W \quad \frac{\Gamma, !A, !A, \Delta \to C}{\Gamma, !A, \Delta \to C}\ !C \quad \frac{\Gamma, B, !A, \Delta \to C}{\Gamma, !A, B, \Delta \to C} \ \frac{\Gamma, !A, B, \Delta \to C}{\Gamma, B, !A, \Delta \to C}\ !P$$

The structural rules $!W$, $!C$, $!P$ are called weakening, contraction, and permutation, respectively.

It happens that for the "quasi-exponential" construction $\mathbf{1} \cap A$ all these rules hold in Lε-models.

Proposition 5 *The following rules are admissible on the class of Lε-models, that is, if their premise is true in an Lε-model $\mathcal{M}$, then so is the conclusion.*

$$\frac{\Gamma, A, \Delta \to C}{\Gamma, \mathbf{1} \cap A, \Delta \to C}\ (\mathbf{1}\cap)L \qquad \frac{\mathbf{1} \cap A_1, \ldots, \mathbf{1} \cap A_n \to B}{\mathbf{1} \cap A_1, \ldots, \mathbf{1} \cap A_n \to \mathbf{1} \cap B}\ (\mathbf{1}\cap)R$$

$$\frac{\Gamma, \Delta \to C}{\Gamma, \mathbf{1} \cap A, \Delta \to C}\ (\mathbf{1}\cap)W \qquad \frac{\Gamma, \mathbf{1} \cap A, \mathbf{1} \cap A, \Delta \to C}{\Gamma, \mathbf{1} \cap A, \Delta \to C}\ (\mathbf{1}\cap)C$$

$$\frac{\Gamma, \mathbf{1} \cap A, B, \Delta \to C}{\Gamma, B, \mathbf{1} \cap A, \Delta \to C} \quad \frac{\Gamma, B, \mathbf{1} \cap A, \Delta \to C}{\Gamma, \mathbf{1} \cap A, B, \Delta \to C} \; (\mathbf{1}\cap)P$$

Proof. The first three rules, $(\mathbf{1}\cap)L$, $(\mathbf{1}\cap)R$, and $(\mathbf{1}\cap)W$ are in fact derivable in the $\mathbf{MALC}^{\varepsilon}$ with $\mathbf{1}$. Due to strong soundness, this makes them admissible in Lε-models. Indeed, $(\mathbf{1}\cap)L$ is an instance of $\cap L_2$; for $(\mathbf{1}\cap)W$ we take a combination of $\cap L_1$ and $\mathbf{1}L$. Finally, the conclusion of $(\mathbf{1}\cap)R$ is derived by $\cap R$ from the premise of this new rule and the sequent $\mathbf{1} \cap A_1, \ldots, \mathbf{1} \cap A_n \to \mathbf{1}$. The latter is derivable from the axiom $\mathbf{1}R$ by applications of $\cap L_1$ and $\mathbf{1}L$.

The remaining rules $(\mathbf{1}\cap)C$ and $(\mathbf{1}\cap)P$ can be derived by Cut and $\cdot L$ with the following sequents:

$$\mathbf{1} \cap A \to (\mathbf{1} \cap A) \cdot (\mathbf{1} \cap A); \quad B \cdot (\mathbf{1} \cap A) \to (\mathbf{1} \cap A) \cdot B; \quad (\mathbf{1} \cap A) \cdot B \to B \cdot (\mathbf{1} \cap A)$$

Thus, it is sufficient to show that these sequents are true in all Lε-models.

This follows from the fact that $\mathsf{w}(\mathbf{1} \cap A) \subseteq \mathsf{w}(\mathbf{1}) = \{\varepsilon\}$, and therefore $\mathsf{w}(\mathbf{1} \cap A)$ is either $\varnothing$ or $\{\varepsilon\}$. In both cases we have $\mathsf{w}(\mathbf{1} \cap A) = \mathsf{w}(\mathbf{1} \cap A) \cdot \mathsf{w}(\mathbf{1} \cap A)$ and $\mathsf{w}(B) \cdot \mathsf{w}(\mathbf{1} \cap A) = \mathsf{w}(\mathbf{1} \cap A) \cdot \mathsf{w}(B)$ (for any $\mathsf{w}(B)$). □

We finish this section and the whole survey by conjecturing an even more interesting complexity result, which could arise in the presence of Kleene star. Here we have a Π_1^1-completeness result for $\mathbf{ACT}_\omega$ extended with the exponential modality [26], so we anticipate the same complexity in the presence of the unit.

The upper bound here follows from the fact that the quantifier over all Lε-models is a second-order arithmetical quantifier, while checking validity of a sequent in a model is expressible in first-order arithmetic.

For the interesting lower bound, one has to modify Kozen's result [20] that reasoning from finite sets of hypotheses in $*$-continuous Kleene algebras is Π_1^1-complete. If we look into Kozen's proof, we see that his encoding uses only $\cdot$ and *. The encoding represents well-foundedness of relations on natural numbers, given by Turing machines. The idea is to combine this idea with Buszkowski's [4] encoding of type-0 grammars (which are equivalent to Turing machines) in the $\backslash$-only fragment of $\mathbf{L}^{\varepsilon}$. If one manages to perform this, the result would be Π_1^1-completeness of reasoning from hypotheses (in $\mathbf{ACT}_\omega$) in the language involving only $\backslash$ and the iterative division $^*\backslash$, instead of Kleene star.

For this fragment, we have strong completeness w.r.t. Lε-models (Theorem 3). Using strong completeness, we can translate derivability from hypotheses in $\mathbf{ACT}_\omega$ into semantic entailment on the class of Lε-models, and then, by Lemma 2, to validity of sequents where hypotheses are encoded using the $\mathbf{1} \cap A$ construction.

An accurate realisation of the aforementioned idea is performed in the new paper [17], which was written after the present paper had been submitted. Thus, we get a huge complexity gap between the language of $\backslash$, $^*\backslash$, and $\cap$ and its extension with the "tiny unit constant" $\mathbf{1}$: Π_1^0 vs. Π_1^1.

References

1. Andréka H, Mikulás S (1994) Lambek calculus and its relational semantics: completeness and incompleteness. J Logic Lang Inf 3(1):1–37. https://doi.org/10.1007/BF01066355
2. Andreoli JM (1992) Logic programming with focusing proofs in linear logic. J Logic Comput 2(3):297–347. https://doi.org/10.1093/logcom/2.3.297
3. Buszkowski W (1982) Compatibility of a categorial grammar with an associated category system. Zeitschrift für Mathematische Logik und Grundlagen der Mathematik 28:229–237. https://doi.org/10.1002/malq.19820281407
4. Buszkowski W (1982) Some decision problems in the theory of syntactic categories. Zeitschrift für Mathematische Logik und Grundlagen der Mathematik 28:539–548. https://doi.org/10.1002/malq.19820283308
5. Buszkowski W (1986) Completeness results for Lambek syntactic calculus. Zeitschrift für Mathematische Logik und Grundlagen der Mathematik 32:13–28. https://doi.org/10.1002/malq.19860320104
6. Buszkowski W (2006) On the complexity of the equational theory of relational action algebras. In: Schmidt RA (ed) Relations and Kleene algebra in computer science, RelMiCS/AKA 2006. Lecture notes in computer science, vol 4136. Springer, Berlin, Heidelberg, pp 106–119. https://doi.org/10.1007/11828563_7
7. Buszkowski W (2007) On action logic: equational theories of action algebras. J Logic Comput 17(1):199–217. https://doi.org/10.1093/logcom/exl036
8. Buszkowski W (2010) Lambek calculus and substructural logics. Linguist Anal 36(1–4):15–48. http://buszko.home.amu.edu.pl/LCaSLpr.pdf
9. Buszkowski W, Palka E (2008) Infinitary action logic: complexity, models and grammars. Stud Logica 89(1):1–18. https://doi.org/10.1007/s11225-008-9116-7
10. Clark A (2015) The syntactic concept lattice: another algebraic theory of the context-free languages? J Logic Comput **25**(5):1203–1229. https://doi.org/10.1093/logcom/ext037
11. Das A, Pous D (2018) Non-wellfounded proof theory for (Kleene + action) (algebras + lattices). In: Ghica D, Jung A (eds) 27th EACSL Annual conference on computer science logic, CSL 2018. Leibniz International proceedings in informatics, vol 119. Schloss Dagstuhl — Leibniz-Zentrum für Informatik, pp 19:1–19:18. https://doi.org/10.4230/LIPIcs.CSL.2018.19
12. Girard JY (1987) Linear logic. Theor Comput Sci 50(1):1–101. https://doi.org/10.1016/0304-3975(87)90045-4
13. de Groote P (2005) On the expressive power of the Lambek calculus extended with a structural modality. In: Language and grammar. CSLI Lecture Notes, vol 168. Stanford University, pp 95–111
14. Kanazawa M (1992) The Lambek calculus enriched with additional connectives. J Logic Lang Inf 1(2):141–171. https://doi.org/10.1007/BF00171695
15. Kanovich M, Kuznetsov S, Scedrov A (2016) Undecidability of the Lambek calculus with a relevant modality. In: Foret A, Morrill G, Muskens R, Osswald R, Pogodalla S (eds) FG 2015, FG 2016: Formal grammar. Lecture notes in computer science, vol 9804. Springer, pp 240–256. https://doi.org/10.1007/978-3-662-53042-9_14
16. Kanovich M, Kuznetsov S, Scedrov A (2022) Language models for some extensions of the Lambek calculus. Inf Comput 287:104760. https://doi.org/10.1016/j.ic.2021.104760
17. Kanovich M, Kuznetsov SL, Scedrov A (2026) Complexity of equational theories for relational and language action lattices. In: Fahrenberg U, Fussner W, Santocanale L (eds) Relational and algebraic methods in computer science. RAMICS 2026. Lecture notes in computer science, vol 16526. Springer, pp 208–228. https://doi.org/10.1007/978-3-032-22469-9_12
18. Kozak M (2009) Distributive full Lambek calculus has the finite model property. Stud Logica 91:201–216. https://doi.org/10.1007/s11225-009-9172-7
19. Kozen D (1994) On action algebras. In: van Eijck J, Visser A (eds) Logic and information flow. MIT Press, pp 78–88
20. Kozen D (2002) On the complexity of reasoning in Kleene algebra. Inf Comput 179(2). https://doi.org/10.1006/inco.2001.2960

21. Kuznetsov S (2018) $*$-continuity vs. induction: divide and conquer. In: Proceedings of advances in modal logic 2018. Advances in modal logic, vol 12. College Publications, London, pp 493–510. https://www.aiml.net/volumes/volume12/Kuznetsov.pdf
22. Kuznetsov SL (2014) Trivalent logics arising from L-models for the Lambek calculus with constants. J Appl Non-Class Logics 14(1–2):132–137. https://doi.org/10.1080/11663081.2014.911522
23. Kuznetsov SL (2024) Strong conservativity and completeness for fragments of infinitary action logic. Siberian Electron Math Rep 21(2):789–809. http://semr.math.nsc.ru/v21/n2/p0789-0809.pdf
24. Kuznetsov SL (2024) Syntactic concept lattice models for infinitary action logic. In: Metcalfe G, Studer T, de Queiroz R (eds) Logic, language, information, and computation, 30th International workshop, WoLLIC 2024. Lecture notes in computer science, vol 14672. Springer, pp 93–107. https://doi.org/10.1007/978-3-031-62687-6_7
25. Kuznetsov SL, Ryzhkova NS (2020) A restricted fragment of the Lambek calculus with iteration and intersection operations. Algebra Logic 59(2):129–146. https://doi.org/10.1007/s10469-020-09586-9
26. Kuznetsov SL, Speranski SO (2022) Infinitary action logic with exponentiation. Ann Pure Appl Logic 173(2):103057. https://doi.org/10.1016/j.apal.2021.103057
27. Lambek J (1958) The mathematics of sentence structure. Am Math Monthly 65:154–170. https://doi.org/10.1080/00029890.1958.11989160
28. Lincoln P, Mitchell J, Scedrov A, Shankar N (1992) Decision problems for propositional linear logic. Ann Pure Appl Log 56(1–3):239–311. https://doi.org/10.1016/0168-0072(92)90075-B
29. Moot R, Retoré C (2012) The logic of categorial grammars: a deductive account of natural language syntax and semantics. Lecture notes in computer science, vol 6850. Springer, Berlin. https://doi.org/10.1007/978-3-642-31555-8
30. Ono H, Komori Y (1985) Logics without the contraction rule. J Symb Log 50(1):169–201. https://doi.org/10.2307/2273798
31. Palka E (2007) An infinitary sequent system for the equational theory of $*$-continuous action lattices. Fundam Inform 78(2):295–309. https://doi.org/10.3233/FUN-2007-78206
32. Pentus M (1994) The conjoinability relation in Lambek calculus and linear logic. J Logic Lang Inf 3(2):121–140. https://doi.org/10.1007/BF01110612
33. Pentus M (1995) Models for the Lambek calculus. Ann Pure Appl Log 75(1–2):179–213. https://doi.org/10.1016/0168-0072(94)00063-9
34. Pentus M (1998) Free monoid completeness of the Lambek calculus allowing empty premises. In: Larrazabal JM, Lascar D, Mints G (eds) Logic Colloquium 1996. Lecture notes in logic, vol 12. Cambridge University Press, pp 171–209. https://doi.org/10.1017/9781316716816.008
35. Pratt V (1991) Action logic and pure induction. In: van Eijck J (ed) JELIA 1990: Logics in AI. Lecture notes in artificial intelligence, vol 478. Springer, pp 97–120. https://doi.org/10.1007/BFb0018436
36. Sedlár I (2020) Iterative division in the distributive full non-associative Lambek calculus. In: Soares Barbosa L, Baltag A (eds) Dynamic logic. New trends and applications. DaLí 2019. Lecture notes in computer science, vol 12005. Springer, pp 141–154. https://doi.org/10.1007/978-3-030-38808-9_9
37. Wurm C (2017) Language-theoretic and finite relation models for the (full) Lambek calculus. J Logic Lang Inf 26(2):179–214. https://doi.org/10.1007/s10849-017-9249-z

On Algebras and Phase Spaces for Linear Logics

Wojciech Buszkowski

Abstract We study algebras and phase spaces for linear logics, including non-associative, non-commutative, obeying Lambek's restriction, and others. We prove some representation theorems: every algebra from the class corresponding to the given logic can be embedded in the algebra of facts of some phase space, even based on a free groupoid. Furthermore, the complete algebras are isomorphic to the algebras of facts of the corresponding phase spaces. At the end, we find some interpretation of cyclic linear negation in terms of syntactic concept lattices.

Keywords Residuated groupoid · Residuated semigroup · Involutive algebra · Cyclic algebra · Phase space · Lattice · Complete lattice · Linear logic

1 Introduction

Phase space semantics for Propositional Linear Logic PLL was introduced by Girard [9], adjusted to Cyclic MALL by Yetter [18] and to Non-commutative MALL by Abrusci [1]; MALL means Multiplicative-Additive Linear Logic, i.e., PLL without exponentials. Phase spaces, however, were not systematically studied for themselves. They were used in proofs of completeness and finite model property as well as in an algebraic proof of cut elimination [10]. Most works in linear logics focused on proof-theoretic methods. Although a closely related research in substructural logics extensively applied algebraic tools, phase spaces were replaced by more general models, e.g., residuated frames [7, 8].

Algebras for Non-commutative MALL admit two negations $^{\sim}, ^{-}$, which are order-reversing and such that $a^{\sim -} = a = a^{- \sim}$. Algebras for Cyclic MALL admit one negation $\sim$, which is order-reversing and such that $a^{\sim\sim} = a$. One also assumes a contraposition law. We refer to algebras with two negations as *involutive* and to those with one negation as *cyclic*.

W. Buszkowski (✉)
Faculty of Mathematics and Computer Science, Adam Mickiewicz University, Poznań, Poland
e-mail: buszko@amu.edu.pl

R. Loukanova (ed.), *Computational Linguistics, Information, Reasoning, and AI 2024*, Studies in Computational Intelligence 1280,
https://doi.org/10.1007/978-3-032-21143-9_3

The author believes that phase spaces deserve a more thorough study. They are simple algebraic structures $(M, \cdot, O)$ such that $(M, \cdot)$ is a monoid, and $O \subseteq M$ satisfies some postulates. This works for associative logics with multiplicative constants. For associative logics obeying Lambek's restriction (sequent systems are restricted to sequents with nonempty antecedents), corresponding to algebras without unit elements for tensor and par, a monoid is replaced by a semigroup, and for non-associative logics by a (unital) groupoid. For logics, obeying Lambek's restriction it is natural to consider relational phase spaces $(M, \cdot, R)$, where R is a binary relation on M. Relational phase spaces are a useful technical tool for studying phase spaces.

Although non-associative linear logics are not often studied in the literature on linear logics, their intuitionistic fragments, e.g., Non-associative Lambek Calculus, are important in the area of categorial grammars; see [3, 13]. Recently there appeared several papers on intuitionistic non-associative linear logics, enriched with modalities, studying their proof-theoretic and computational aspects; see e.g., [2].

The present paper continues Buszkowski [4, 5], where non-associative phase spaces were used to establish different properties of Non-associative Lambek Calculus and its extensions with cyclic and involutive negations, e.g., algebraic proofs of cut elimination, finite model property of logics enriched with non-logical axioms (assumptions), strong conservativity of logics with negation(s) over their negation-free fragments, and some decidability and complexity results. Płaczek [14] obtained some related results for richer logics.

Here we focus on relations between algebras and phase spaces for different linear logics. Proof-theoretic methods are not used. The main results are as follows. Theorem 1 states that every involutive (resp. cyclic) algebra can be embedded in the algebra of facts in some involutive (resp. cyclic) relational phase space. Furthermore, complete algebras are isomorphic to the resulting algebras of facts. If the source algebra is associative (resp. commutative, unital), then the target algebra of facts is so. Involutive and cyclic phase spaces are defined by special postulates. Theorem 2 states that two relational phase spaces such that one is a homomorphic image of the other yield the same algebra of facts (up to isomorphism). Theorem 3 states that every involutive (resp. cyclic) algebra can be embedded in the algebra of facts in some involutive (resp. cyclic) phase space, based on a free groupoid. Again, complete algebras are isomorphic to the resulting algebras of facts. The preservation properties hold, as above.

In Sect. 2 we survey algebras corresponding to linear logics and their fragments. In Sect. 3 we consider phase spaces and prove the main results. Section 4 provides an interpretation of cyclic negation in terms of syntactic concept lattices. The latter structures were introduced by Clark [6] as models of the family of syntactic categories of a language. Section 5 contains some conclusions.

Our notation is not the standard notation in the literature on linear logic. Tensor is denoted by $\odot$, par by $\bullet$, additive operations by $\sqcup, \sqcap$, multiplicative constants by $1, 0$, additive constants by $\top, \bot$, implications by $\backslash, /$, and negations by $^{\sim}, ^{-}$. So our notation is similar to that used in the literature on Lambek calculi and substructural logics; see [8]. The operation $\odot$ is called multiplication. (It is called *product* in many papers on Lambek calculi, but this terminology is incompatible with the meaning of product in type theories and category theory: product corresponds to $\sqcap$.) We also write O instead of $\bot^{\bullet}$; the set O is the element 0 in the algebra of facts, if it is unital.

2 Algebras

We define algebras, corresponding to Lambek calculi (residuated groupoids and semigroups) and substructural logics (algebras with lattice operations).

A residuated groupoid is an ordered algebra $\mathbf{G} = (G, \odot, \backslash, /, \leq)$ such that $(G, \leq)$ is a poset and $\odot, \backslash, /$ are binary operations on G, satisfying the residuation laws:

$$\text{(RES) } a \odot b \leq c \text{ iff } b \leq a\backslash c \text{ iff } a \leq c/b\,, \quad \text{for all } a, b, c \in G\,.$$

We refer to $\odot$ as *multiplication* and to $\backslash, /$ as *the right residuation* and *the left residuation*, respectively, of $\odot$ (they are also called implications).

If $\odot$ is associative, then $\mathbf{G}$ is called *a residuated semigroup*. If $1 \in G$ satisfies $1 \odot a = a = a \odot 1$, then $\mathbf{G}$ is said to be *unital*; this term will also be applied to other algebras with multiplication and 1. Unital semigroups are called *monoids*, hence unital residuated semigroups are also called residuated monoids.

An (ordered) algebra is said to be *bounded*, if it contains the least element $\bot$ and the greatest element $\top$. A residuated groupoid with $\bot$ is bounded, since $\bot\backslash\bot$ is the greatest element.

Below we show some laws valid in these algebras. Those labeled by (RGi) are valid in residuated groupoids, by (bRGi) in bounded residuated groupoids, by (URGi) in unital residuated groupoids, and by (RSi) in residuated semigroups.

(RG1) $a \odot (a\backslash b) \leq b$, $(a/b) \odot b \leq a$
(RG2) $b \leq a\backslash(a \odot b)$, $a \leq (a \odot b)/b$
(RG3) $a \leq b \Rightarrow c \odot a \leq c \odot b \wedge a \odot c \leq b \odot c$
(RG4) $a \leq b \Rightarrow c\backslash a \leq c\backslash b \wedge a/c \leq b/c$
(RG5) $a \leq b \Rightarrow b\backslash c \leq a\backslash c \wedge c/b \leq c/a$
(RG6) $a \leq (b/a)\backslash b$, $a \leq b/(a\backslash b)$
(bRG1) $\bot \odot a = \bot = a \odot \bot$
(bRG2) $\bot\backslash a = \top = a/\bot$
(URG1) $a \leq b$ iff $1 \leq a\backslash b$ iff $1 \leq b/a$
(URG2) $1\backslash a = a = a/1$
(URG3) $(a/a)\backslash a = a = a/(a\backslash a)$
(RS1) $(a\backslash b)/c = a\backslash(b/c)$
(RS2) $(a \odot b)\backslash c = b\backslash(a\backslash c)$, $a/(b \odot c) = (a/c)/b$

(URG3) is not valid in residuated groupoids; we only have $a \leq (a/a)\backslash a$ and $a \leq a/(a\backslash a)$, by (RG6). Therefore the equational theory of residuated groupoids is properly contained in the equational theory of unital residuated groupoids in the language without 1. This remains true for all algebras, considered in this paper.

Non-associative Lambek Calculus can be seen as the atomic theory of residuated groupoids: it provides all laws $s \leq t$ valid in this class, where s, t are terms in language $\odot, \backslash, /$. Lambek Calculus corresponds to residuated semigroups. Lambek Calculus was introduced by Lambek [11], under the name Syntactic Calculus, as a logic for categorial grammars, and its non-associative version by Lambek [12]. One also considers versions of these logics with $1, \top, \bot$.

A residuated groupoid $\mathbf{G}$ is said to be *commutative*, if $\odot$ is commutative. Then $a\backslash b = b/a$ for all $a, b \in G$, and one writes $a \rightarrow b$ for both. The laws listed above look differently. We write two examples.

(RG6c) $a \leq (a \rightarrow b) \rightarrow b$
(RS1c) $c \rightarrow (a \rightarrow b) = a \rightarrow (c \rightarrow b)$

Commutative Lambek Calculus is used in type-theoretic semantics for Lambek calculi and categorial grammars; see van Benthem [15].

An algebra $(G, \odot, \backslash, /, \sqcup, \sqcap)$ is called *a residuated lattive-ordered groupoid* (shortly: r.l. groupoid), if $(G, \sqcup, \sqcap)$ is a lattice and $(G, \odot, \backslash, /, \leq)$ is a residuated groupoid, where $\leq$ is the lattice ordering: $a \leq b \Leftrightarrow a \sqcup b = b$. A r.l. monoid (i.e., a unital r.l. semigroup) is called *a residuated lattice*. The following laws are valid in r.l. groupoids.

(LRG1) $(a \sqcup b) \odot c = (a \odot c) \sqcup (b \odot c), a \odot (b \sqcup c) = (a \odot b) \sqcup (a \odot c)$
(LRG2) $a\backslash(b \sqcap c) = (a\backslash b) \sqcap (a\backslash c), (a \sqcap b)/c = (a/c) \sqcap (b/c)$
(LRG3) $(a \sqcup b)\backslash c = (a\backslash c) \sqcap (b\backslash c), a/(b \sqcup c) = (a/b) \sqcap (a/c)$

In fact, these laws can be extended for infinite joins and meets. (LRG1) extends to:

(LRG1') $(\bigvee_{i \in I} a_i) \odot b = \bigvee_{i \in I}(a_i \odot b), b \odot \bigvee_{i \in I} a_i = \bigvee_{i \in I}(b \odot a_i)$,

if $\bigvee_{i \in I} a_i$ exists. The formulations of (LRG2') and (LRG3') are left to the reader; one must assume that the bounds on the left-hand side exist.

Residuated lattices are algebraic models of substructural logics. Full Lambek Calculus can be identified with the equational theory of residuated lattices [8].

We say that an algebra (with lattice operations) is *complete*, if its lattice reduct is a complete lattice. If $\mathbf{G}$ is a complete residuated lattice, then its $(\odot, \sqcup, 1)$−reduct is a quantale, i.e., a complete join-semilattice such that $(G, \odot, 1)$ is a monoid and (LRG1') are valid. Every quantale $\mathbf{Q}$ can be expanded to a complete residuated lattice; one defines $a\backslash b = \bigvee\{x \in Q : a \odot x \leq b\}$, $a/b = \bigvee\{x \in Q : x \odot b \leq a\}$. Yetter [18] used some special quantales as algebraic models of Cyclic MALL. In what follows we do not refer to quantales explicitly.

Now, we define some algebras corresponding to linear logics. *An involutive r.l. groupoid* is an r.l. groupoid with two unary operations $^{\sim}, ^{-}$, which are order reversing and satisfy the following laws for all elements a, b:

$$\text{(DN) } a^{-\sim} = a = a^{\sim-},$$

$$\text{(CONT) } a^{\sim}/b = a\backslash b^{-}.$$

(DN) are the double negation laws. (CONT) is an intuitionistic-style contraposition law. One derives other contraposition laws $a/b = a^{-}\backslash b^{-}$, $a\backslash b = a^{\sim}/b^{\sim}$ and the following equivalences.

$$\text{(CONT')}\ a \odot b \leq c \text{ iff } b \odot c^{\sim} \leq a^{\sim} \text{ iff } c^{-} \odot a \leq b^{-}$$

A cyclic involutive r.l. groupoid is an involutive r.l. groupoid such that $a^{\sim} = a^{-}$ for all elements a (we denote the operation by $^{\sim}$). Later on we write 'cyclic' for 'cyclic involutive'. In cyclic algebras (CONT) admits the form: $a^{\sim}/b = a\backslash b^{\sim}$.

In involutive r.l. unital groupoids $1^{\sim} = 1^{\sim}/1 = 1\backslash 1^{-} = 1^{-}$. One defines $0 = 1^{\sim}$ and obtains $a^{\sim} = a\backslash 0$, $a^{-} = 0/a$. If $\odot$ is associative, admits $1, 0$, and negations are defined by the latter equations, then (CONT) follows from (RS1) and can be omitted in the definition.

We write some laws valid in involutive r.l. groupoids.

(IRG1) $(a^{-} \odot b^{-})^{\sim} = (a^{\sim} \odot b^{\sim})^{-}$
(IRG2) $(a \sqcup b)^{\sim} = a^{\sim} \sqcap b^{\sim}$, $(a \sqcup b)^{-} = a^{-} \sqcap b^{-}$
(IRG3) $(a \sqcap b)^{\sim} = a^{\sim} \sqcup b^{\sim}$, $(a \sqcap b)^{-} = a^{-} \sqcup b^{-}$
(bIRG1) $\bot^{\sim} = \top = \bot^{-}$, $\top^{\sim} = \bot = \top^{-}$

To prove (IRG1) one shows: $x \leq (a^{-} \odot b^{-})^{\sim} \Leftrightarrow x \leq (a^{\sim} \odot b^{\sim})^{-}$, using order reversing, (DN) and (CONT') several times. The remaining equations are easy to prove. One defines a new operation $\bullet$: $a \bullet b = (b^{-} \odot a^{-})^{\sim}$, called *par* in linear logics. In involutive r.l. unital groupoids $0 \bullet a = a = a \bullet 0$. The following laws show that some operations are definable in terms of others.

(IRG4) $a\backslash b = a^{\sim} \bullet b$, $a/b = a \bullet b^{-}$
(IRG5) $a \odot b = (b\backslash a^{\sim})^{-} = (b^{-}/a)^{\sim}$

Involutive r.l. monoids are algebraic models of Non-commutative MALL, cyclic r.l. monoids of Cyclic MALL and commutative cyclic r.l. monoids of MALL. We note that every commutative involutive r.l. unital groupoid must be cyclic: $a^{\sim} = a\backslash 0 = 0/a = a^{-}$. There exist non-cyclic involutive commutative r.l. semigroups. For example, we take the set of rational numbers $\mathbb{Q}$ with the natural order, extended with $\bot, \top$. We define $a \odot b = \bot$ for all elements a, b, which yields $a\backslash b = b/a = \top$. Two different negations $^{\sim}, ^{-}$ can be defined as follows: $a^{\sim} = -2a$, $a^{-} = -\frac{a}{2}$ for $a \in \mathbb{Q}$, $\bot^{\sim} = \bot^{-} = \top$ and $\top^{\sim} = \top^{-} = \bot$. Clearly $^{\sim}, ^{-}$ reverse ordering and satisfy (DN), (CONT).

Complete commutative cyclic r.l. monoids are algebraic models of PLL: the completeness is needed to define exponentials by means of infinite meets and joins.

3 Phase Spaces

We need some notions connected with partially ordered sets (posets). Let $(P, \leq)$ be a poset. For $a \in P$, we define $a^{\downarrow} = \{x \in P : x \leq a\}$. A map $\eta : P \mapsto P$ is called *a closure operation* on $(P, \leq)$, if it satisfies: (C1) $x \leq \eta(x)$, (C2) $x \leq y \Rightarrow \eta(x) \leq \eta(y)$, (C3) $\eta(\eta(x)) \leq \eta(x)$, for all $x, y \in P$. We will use this notion for posets $(\mathcal{P}(M), \subseteq)$ only. Let η be a closure operation of $\mathcal{P}(M)$. A set $X \subseteq M$ is said to be $\eta-$closed, if $\eta(X) \subseteq X$ (equivalently $\eta(X) = X$). The family of $\eta-$closed subsets of M is denoted by M_η. This family is closed under infinite meets.

Let $f, g : \mathcal{P}(M) \mapsto \mathcal{P}(M)$. The pair (f, g) is called *a Galois correspondence*, if $X \subseteq f(Y) \Leftrightarrow Y \subseteq g(X)$ for all $X, Y \subseteq M$. It follows that f, g reverse $\subseteq$ and both $f \circ g$ and $g \circ f$ are closure operations on $\mathcal{P}(M)$. For (C3), one uses $f \circ g \circ f = f$, $g \circ f \circ g = g$. Furthermore, X is $f \circ g-$closed (resp. $g \circ f-$closed) if and only if $X = f(Y)$ (resp. $X = g(Y)$) for some $Y \subseteq M$. There hold the following equations:

$$f(\bigcup_{i \in I} X_i) = \bigcap_{i \in I} f(X_i) \text{ and similarly for } g \tag{1}$$

for every family $\{X_i\}_{i \in I}$ of subsets of M.

Let $(M, \cdot)$ be a groupoid. We consider the powerset $\mathcal{P}(M)$. One defines operations $\otimes, \backslash, /$ on $\mathcal{P}(M)$.

$$X \otimes Y = \{a \cdot b : a \in X, b \in Y\} \tag{2}$$

$$X\backslash Y = \{c \in M : \forall_{a \in X} a \cdot c \in Y\}; \quad X/Y = \{c \in M : \forall_{b \in Y} c \cdot b \in X\} \tag{3}$$

One easily shows: $X \otimes Y \subseteq Z$ iff $Y \subseteq X\backslash Z$ iff $X \subseteq Z/Y$. So $\mathcal{P}(M)$ with these operations and $\subseteq$ is a residuated groupoid. With operations $\cup, \cap$ it becomes a (complete) r.l. groupoid; we refer to it as *the powerset algebra* of the groupoid $(M, \cdot)$. If $(M, \cdot, 1)$ is a unital groupoid, this construction yields a unital r.l. groupoid, having $\{1\}$ as the unit for $\otimes$. If $\cdot$ is associative (resp. commutative), then $\otimes$ is associative (resp. commutative). Accordingly, the powerset algebra of a semigroup (resp. monoid) is a r.l. semigroup (resp. residuated lattice).

Let $(M, \cdot)$ be as above. A closure operation η on $\mathcal{P}(M)$ is called *a nucleus*, if it satisfies: (C4) $\eta(X) \otimes \eta(Y) \subseteq \eta(X \otimes Y)$ for all $X, Y \subseteq M$. Assuming (C1)-(C3), (C4) is equivalent to: (C4') for every $X \in M_\eta$ and every $Y \subseteq M$, the sets $Y\backslash X$ and X/Y are in M_η [4]. We only need (C4')$\Rightarrow$(C4) and show it. Assume (C4'). From $X \otimes Y \subseteq \eta(X \otimes Y)$ we obtain $Y \subseteq X\backslash\eta(X \otimes Y)$, hence $\eta(Y) \subseteq X\backslash\eta(X \otimes Y)$, by (C4'). This yields $X \otimes \eta(Y) \subseteq \eta(X \otimes Y)$, hence we obtain (C4) by a similar argument.

By *a relational phase space* we mean a structure $(M, \cdot, R)$ such that $(M, \cdot)$ is a groupoid and $R \subseteq M^2$. One defines the operations $^\sim, ^-$ on $\mathcal{P}(M)$.

$$X^\sim = \{a \in M : \forall_{b \in X} R(b, a)\} \quad X^- = \{a \in M : \forall_{b \in X} R(a, b)\} \tag{4}$$

The pair $^\sim, ^-$ is a Galois correspondence: $X \subseteq Y^\sim$ iff $Y \subseteq X^-$; either condition is equivalent to $\forall_{a \in X} \forall_{b \in Y} R(b, a)$. Accordingly $^\sim$ and $^-$ reverse $\subseteq$ and the maps $\phi_R(X) = X^{-\sim}$ and $\psi_R(X) = X^{\sim-}$ are closure operations on $\mathcal{P}(M)$. Consequently, X is ϕ_R-closed (resp. ψ_R-closed) if and only if $X = Y^\sim$ (resp. $X = Y^-$) for some $Y \subseteq M$.

The following proposition is well-known [8]: if η is a nucleus, then M_η with operations $\odot_\eta, \backslash, /, \cup_\eta, \cap$, where $X \odot_\eta Y = \eta(X \otimes Y)$ and $X \cup_\eta Y = \eta(X \cup Y)$, is a complete r.l. groupoid ($\backslash, /$ are the operations on $\mathcal{P}(M)$, restricted to M_η).

A relational phase space $(M, \cdot, R)$ is said to be *involutive*, if $\phi_R = \psi_R$ and

(RPS1) $R(a \cdot b, c)$ iff $R(a, b \cdot c)$, for all $a, b, c \in M$,

and *cyclic*, if it satisfies (RPS1) and

(RPS2) if $R(a, b)$ then $R(b, a)$, for all $a, b \in M$.

Clearly (RPS2) means that R is symmetrical. Assuming (RPS2), (RPS1) can be replaced by: if $R(a \cdot b, c)$ then $R(a, b \cdot c)$. The converse implication follows (below each atomic formula implies the next one, if exists). We write ab for $a \cdot b$, and similarly for other terms.

$$R(a, bc), R(bc, a), R(b, ca), R(ca, b), R(c, ab), R(ab, c)$$

(RPS2) is equivalent to: $X^{\sim} = X^{-}$ for all $X \subseteq M$, which yields $\phi_R = \psi_R$. So every cyclic relational phase space is involutive.

We write M_R for M_{ϕ_R}, $\odot_R$ for $\odot_{\phi_R}$ and $\cup_R$ for $\cup_{\phi_R}$.

Lemma 1 *For any involutive (resp. cyclic) relational phase space* $(M, \cdot, R)$ *, the algebra* $(M_R, \odot_R, \backslash, /, ^{\sim}, ^{-}, \cup_R, \cap)$ *is a complete involutive (resp. cyclic) r.l. groupoid.*

Proof First, we prove the lemma for involutive $(M, \cdot, R)$. We show that ϕ_R satisfies (C4'). Let X be ϕ_R−closed and $Y \subseteq M$. Then $X = Z^{\sim}$ for some $Z \subseteq M$. We show $Y \backslash Z^{\sim} = (Z \otimes Y)^{\sim}$: $a \in Y \backslash Z^{\sim}$ iff for all $y \in Y$, $ya \in Z^{\sim}$ iff for all $z \in Z$, $y \in Y$, $R(z, ya)$ iff for all $z \in Z$, $y \in Y$, $R(zy, a)$, by (RPS1), iff $a \in (Z \otimes Y)^{\sim}$. Consequently $Y \backslash X$ is ϕ_R−closed. Since $\phi_R = \psi_R$, then X is ψ_R−closed, and consequently $X = U^{-}$ for some $U \subseteq M$. We have $U^{-}/Y = (Y \otimes U)^{-}$, hence X/Y is ψ_R−closed and, obviously, ϕ_R−closed.

Therefore ϕ_R is a nucleus, and $(M_R, \odot_R, \backslash, /, \cup_R, \cap)$ is a r.l. groupoid. For any $X \subseteq M$, $X^{\sim}$ and X^{-} are ϕ_R−closed, hence M_R is closed under $^{\sim}$ and $^{-}$. If $X \in M_R$, then $X^{-\sim} = \phi_R(X) = X$ and $X^{\sim-} = \psi_R(X) = X$. So (DN) are satisfied. For all $X, Y \subseteq M$, $X^{\sim}/Y = X \backslash Y^{-}$ by (RPS1). This yields (CONT).

Assume that $(M, \cdot, R)$ is cyclic. By the above $(M_R, \odot_R, \backslash, /, ^{\sim}, ^{-}, \cup_R, \cap)$ is an involutive r.l. groupoid. It is cyclic, by (RPS2).

M_R is closed under $\bigcap$, hence the resulting algebra is complete.

In what follows the ϕ_R−closed sets are called *facts*. The algebra $(M_R, \ldots)$ from Lemma 1 is called *the algebra of facts* in the relational phase space $(M, \cdot, R)$ and denoted $F(M, \cdot, R)$.

The notion of a relational phase space generalizes a more standard notion of a phase space. By *a phase space* we mean a structure $(M, \cdot, O)$ such that $(M, \cdot)$ is a groupoid and $O \subseteq M$. One defines a relation R_O.

$$R_O(a, b) \text{ iff } a \cdot b \in O, \text{ for } a, b \in M$$

The operations $^{\sim}, ^{-}$ are defined by (4) with $R = R_O$. Then $X^{\sim} = X \backslash O$ and $X^{-} = O/X$ for all $X \subseteq M$.

A phase space $(M, \cdot, O)$ is said to be *involutive* (resp. *cyclic*), if $(M, \cdot, R_O)$ is involutive (resp. cyclic). We write ϕ_O, ψ_O for ϕ_{R_O}, ψ_{R_O}, M_O for M_{R_O}, and so on. Clearly $(M, \cdot, O)$ is involutive if and only if $\phi_O = \psi_O$ and

(PS1) $(a \cdot b) \cdot c \in O$ iff $a \cdot (b \cdot c) \in O$, for all $a, b, c \in M$.

It is cyclic if and only if it satisfies (PS1) and

(PS2) if $a \cdot b \in O$ then $b \cdot a \in O$, for all $a, b \in M$.

Corollary 1 *Let $(M, \cdot, O)$ be an involutive (resp. cyclic) phase space. Then $(M_O, \odot_O, \backslash, /, ^{\sim}, ^{-}, \cup_O, \cap)$ is an involutive (resp. cyclic) r.l. groupoid.*

We define $F(M, \cdot, O) = F(M, \cdot, R_O)$; this is the algebra of facts in $(M, \cdot, O)$. If $\cdot$ is associative, then (PS1) is true. If $\cdot$ is commutative, then (PS2) is true. So, if $(M, \cdot)$ is a semigroup, then $F(M, \cdot, O)$ is involutive, for any $O \subseteq M$, satisfying $\phi_O = \psi_O$. If $(M, \cdot)$ is a commutative semigroup, then $F(M, \cdot, O)$ is cyclic, for any $O \subseteq M$.

Remark 1 If $\cdot$ admits 1, then $O = \{1\} \backslash O = \{1\}^{\sim}$, hence $O \in M_O$. Otherwise, O need not be ϕ_O−closed. Let us consider the phase space $(\Sigma^+, \cdot, O)$ such that Σ^+ is the set of nonempty finite sequences of elements of a nonempty set Σ, $\cdot$ is the concatenation of sequences from Σ^+, and O is the set of all sequences of length 1. (PS1) is true, since both sides of the equivalence are false. (PS2) is true, since the left hand side of the implication is false. So this phase space is cyclic. We have $\emptyset^{\sim} = \Sigma^+$ and $X^{\sim} = \emptyset$, for any nonempty $X \subseteq \Sigma^+$. Consequently $\mathcal{P}_O(\Sigma^+) = \{\emptyset, \Sigma^+\}$ and $O \notin \mathcal{P}_O(\Sigma^+)$.

Remark 2 Not every relational phase space is of the form $(M, \cdot, R_O)$, for some phase space $(M, \cdot, O)$. We construct a relational phase space $(M, \cdot, R)$ such that $R \neq R_O$ for any $O \subseteq M$. Let M be the set of all pairs of positive integers. For $a, b \in M$, we define $a + b$ pointwise: $(a_1, a_2) + (b_1, b_2) = (a_1 + b_1, a_2 + b_2)$. We define: $R(a, b)$ iff neither a, nor b is of the form $x + y$ for $x, y \in M$. Clearly $(M, +, R)$ is a cyclic relational phase space. Suppose $R = R_O$ for some $O \subseteq M$. We have $R((1, 2), (2, 1))$, hence $(3, 3) \in O$. Since $(3, 3) = (1, 1) + (2, 2)$, then $R((1, 1), (2, 2))$. This contradicts the definition of R, since $(2, 2) = (1, 1) + (1, 1)$. For further purposes, we look more closely at the algebra of facts. Let $A = \{(m, n) \in M : m = 1 \vee n = 1\}$. We have $X^{\sim} = A$ for any nonempty $X \subseteq A$, $X^{\sim} = \emptyset$ for other nonempty $X \subseteq M$, and $\emptyset^{\sim} = M$. So $M_R = \{\emptyset, A, M\}$ and $X \odot_R Y = M$ for all nonempty $X, Y \in M_R$.

The phase space from Remark 2 is associative and commutative, i.e., the groupoid operation has these properties. If $(M, \cdot, R)$ is associative (resp. commutative), then the resulting algebra of facts is associative (resp. commutative). For the operation $\otimes$, and consequently $\odot_R$, is so. To show the associativity of $\odot_R$, one uses $\eta(\eta(X) \otimes \eta(Y)) = \eta(X \otimes Y)$ for all $X, Y \subseteq M$ (this holds for every nucleus η).

A (relational) phase space is said to be *unital*, if $\cdot$ admits a unit element 1. We note that the spaces from Remarks 1 and 2 are not unital. If $(M, \cdot, R)$ is unital and satisfies (RPS1), then $R = R_O$ for $O = \{a \in M : R(1, a)\}$: O satisfies (PS1). For any free groupoid $(M, \cdot)$ and every relation $R \subseteq M^2$, we have $R = R_O$ for $O = \{a \cdot b : R(a, b)\}$; O satisfies (PS1), if R satisfies (RPS1). Consequently, all unital relational phase spaces, satisfying (RPS1), can be obtained from some phase spaces, satisfying (PS1), and similarly for the relational phase spaces based on free groupoids.

If $(M, \cdot, 1)$ is a unital groupoid and η is a nucleus on $(M, \cdot)$, then $\eta(\{1\})$ is the unit for $\odot_\eta$. Consequently, if $(M, \cdot, R)$ is an involutive unital relational phase space, then $\phi_R(\{1\})$ is the unit for $\odot_R$.

Let $\mathbf{G} = (G, \odot, \backslash, /, ^\sim, ^-, \sqcup, \sqcap)$ be an involutive r.l. groupoid. We define a relation $R_\mathbf{G} \subseteq G^2$, by setting: $R_\mathbf{G}(a, b)$ iff $a \leq b^-$ (equivalently: $b \leq a^\sim$). We note that, if $\mathbf{G}$ is unital, then $R_\mathbf{G}(a, b)$ iff $a \odot b \leq 0$, and consequently, if $a \odot b = c \odot d$, then: $R_\mathbf{G}(a, b)$ iff $R_\mathbf{G}(c, d)$. The latter implication, however, need not hold for non-unital (even cyclic) algebras, e.g., the algebra of facts from Remark 2: $A \odot_R A = M = A \odot_R M$, $A \subseteq A^\sim$, but $M \not\subseteq A^\sim$.

The following property immediately follows from the definition of $R_\mathbf{G}$.

(P1) If $R_\mathbf{G}(a, b)$, $a' \leq a$ and $b' \leq b$, then $R_\mathbf{G}(a', b')$.

The following lemma was proved in [1] for associative unital phase spaces, and the proof essentially relies on associativity and the presence of 1. Our lemma is true for arbitrary relational phase spaces.

Lemma 2 *Let $(M, \cdot, R)$ be a relational phase space. Then $\phi_R = \psi_R$ if and only if, for any $a \in M$, the set $\{a\}^-$ is ϕ_R−closed and the set $\{a\}^\sim$ is ψ_R−closed.*

Proof The 'only if' part is obvious. We prove the 'if' part. Assume the right-hand side of the equivalence. We show that every ψ_R−closed set is ϕ_R−closed. We use the following equalities, an instance of (1).

$$(\bigcup_{i \in I} X_i)^\sim = \bigcap_{i \in I} X_i^\sim \quad (\bigcup_{i \in I} X_i)^- = \bigcap_{i \in I} X_i^-$$

Let X be ψ_R−closed. Then $X = Y^-$ for some $Y \subseteq M$. Since $Y = \bigcup_{a \in Y}\{a\}$, then Y^- equals the intersection of a family of sets $\{a\}^-$, which are ϕ_R−closed. Therefore X is ϕ_R−closed. In a similar way one shows that every ϕ_R−closed set is ψ_R−closed. This yields $\phi_R = \psi_R$, since $\eta(X)$ equals the intersection of all η−closed sets, containing X, for any closure operation η.

Theorem 1 *For every involutive (resp. cyclic) r.l. groupoid $\mathbf{G}$, the relational phase space $(G, \odot, R_\mathbf{G})$ is involutive (resp. cyclic), and there exists an embedding of $\mathbf{G}$ in $F(G, \odot, R_\mathbf{G})$, preserving $\bot, \top, 1$, if they belong to $\mathbf{G}$. Furthermore, if $\mathbf{G}$ is complete, then $\mathbf{G}$ is isomorphic to $F(G, \odot, R_\mathbf{G})$.*

Proof Let **G** be an involutive r.l. groupoid. First, we prove that $(G, \odot, R_{\mathbf{G}})$ is involutive. We write R for $R_{\mathbf{G}}$. We have: $x \leq a$ iff $R(x, a^{\sim})$ iff $R(a^{-}, x)$, which yields:

$$a^{\downarrow} = \{a^{\sim}\}^{-} = \{a^{-}\}^{\sim}, \text{ hence } (a^{\sim})^{\downarrow} = \{a\}^{\sim}, (a^{-})^{\downarrow} = \{a\}^{-} \tag{5}$$

for any $a \in G$. Here the inner negations are the operations in **G** and the outer negations are the operations on $\mathcal{P}(G)$, defined by (4). Accordingly, $a^{\downarrow}$ is both ϕ_R–closed and ψ_R–closed. Furthermore, $\{a\}^{-}$ is ϕ_R–closed and $\{a\}^{\sim}$ is ψ_R–closed, for any $a \in G$. By Lemma 2, $\phi_R = \psi_R$. By (CONT') for **G**, $R(a \odot b, c)$ iff $a \odot b \leq c^{-}$ iff $b \odot c \leq a^{\sim}$ iff $R(a, b \odot c)$. So R satisfies (RPS1).

Second, we show that the map $h : G \mapsto M_R$, defined by $h(a) = a^{\downarrow}$ is an embedding of **G** in $F(G, \odot, R)$. We have: $a \leq b$ iff $h(a) \subseteq h(b)$, hence h is an injection. In involutive r.l. groupoids, all operations are definable in terms of $\backslash, ^{\sim}, ^{-}, \sqcap$.

$$a \odot b = (b\backslash a^{\sim})^{-} \quad a/b = a^{-}\backslash b^{-} \quad a \sqcup b = (a^{\sim} \sqcap b^{\sim})^{-} \tag{6}$$

So it suffices to prove that h preserves these four operations. $(a\backslash b)^{\downarrow} = a^{\downarrow}\backslash b^{\downarrow}$ and $(a \sqcap b)^{\downarrow} = a^{\downarrow} \cap b^{\downarrow}$ are valid in r.l. groupoids, hence h preserves $\backslash$ and meet. We have: $x \leq a^{\sim}$ iff $R(a, x)$ iff $\forall_{y \in a^{\downarrow}} R(y, x)$, by (P1), iff $x \in (a^{\downarrow})^{\sim}$. This yields $(a^{\sim})^{\downarrow} = (a^{\downarrow})^{\sim}$. The proof of $(a^{-})^{\downarrow} = (a^{\downarrow})^{-}$ is similar.

Assume $\top \in G$. Since $\top^{\downarrow} = G$, then h preserves $\top$. If $\bot \in G$, then $\top \in G$ and $\bot = \top^{\sim}$. We have $\bot^{\downarrow} = (\top^{\sim})^{\downarrow} = (\top^{\downarrow})^{\sim} = G^{\sim}$, hence h preserves $\bot$ (clearly $G^{\sim}$ is the least fact).

As an immediate consequence of (P1), we obtain: (P2) for all facts X and $a \in G$, if $a \in X$ then $a^{\downarrow} \subseteq X$ (use $X = Y^{\sim}$). Now assume $1 \in G$. From $\{1\} \subseteq 1^{\downarrow}$ we obtain $\phi_R(\{1\}) \subseteq 1^{\downarrow}$, since $1^{\downarrow}$ is a fact. From $1 \in \phi_R(\{1\})$ we obtain $1^{\downarrow} \subseteq \phi_R(\{1\})$, by (P2). So $1^{\downarrow} = \phi_R(\{1\})$, hence h preserves the multiplicative unit.

If **G** is cyclic, then R satisfies (RPS2). We have: $R(a, b)$ iff $a \leq b^{\sim}$ iff $R(b, a)$.

Let **G** be complete. Then h preserves infinite joins and meets. It preserves meets, since $(\bigwedge_{i \in I} a_i)^{\downarrow} = \bigcap_{i \in I} (a_i)^{\downarrow}$ holds in all complete lattices. It preserves joins, by the following argument. The infinite De Morgan law:

$$(\bigvee_{i \in I} a_i)^{-} = \bigwedge_{i \in I} a_i^{-}$$

holds in every complete lattice with operations $^{\sim}, ^{-}$, which reverse ordering and satisfy (DN), and consequently $\bigvee_{i \in I} a_i = (\bigwedge_{i \in I} a_i^{-})^{\sim}$. So h preserves joins, since it preserves meets and $^{\sim}, ^{-}$ We show that h is a surjection. Let $X = Y^{\sim}$ for $Y \subseteq G$. Then $X = \bigcap_{a \in Y} \{a\}^{\sim} = \bigcap_{a \in Y} (a^{\sim})^{\downarrow} = h(\bigwedge_{a \in Y} a^{\sim})$. So h is an isomorphism.

We have shown that the involutive (resp. cyclic) r.l. groupoids coincide (up to isomorphism) with the subalgebras of the algebras of facts in involutive (resp. cyclic) relational phase spaces. Moreover, the complete involutive (resp. cyclic) r.l. groupoids coincide with the algebras of facts in involutive (resp. cyclic) relational phase spaces. Also, if **G** is associative (resp. commutative ,unital), then $F(G, \odot, R_{\mathbf{G}})$ is associative (resp. commutative, unital).

Below we obtain analogous results for phase spaces. We need some auxiliary notions and a lemma.

We need a general lemma on relational structures. Let **M**, **N** be relational structures of the same type (they may contain designated relations, operations and elements. A function $g : M \mapsto N$ is called *a homomorphism* of **M** in **N**, if g preserves the designated operations and elements and the designated relations in the following sense:

$$\text{for all } a_1, \ldots, a_n \in M,\ R^{\mathbf{M}}(a_1, \ldots, a_n) \Leftrightarrow R^{\mathbf{N}}(g(a_1), \ldots, g(a_n)).$$

A homomorphism g of **M** onto **N** is called *an epimorphism*. We consider formulas of the first-order language for this type. We write $t(x_1, \ldots, x_n)$ (resp. $\varphi(x_1, \ldots, x_n)$) for a term (resp. formula) containing no (resp. free) variables different from $x_1, \ldots, x_n$.

Lemma 3 *Let g be an epimorphism of* **M** *onto* **N**. *Then, for every formula $\varphi(x_1, \ldots, x_n)$ without =, the following condition is true:*

$$\textit{for all } a_1, \ldots, a_n \in M, \mathbf{M} \models \varphi[a_1, \ldots, a_n] \Leftrightarrow \mathbf{N} \models \varphi[g(a_1), \ldots, g(a_n)].$$

Proof By induction on t one proves $g(t^{\mathbf{M}}[a_1, \ldots, a_n]) = t^{\mathbf{N}}[g(a_1), \ldots, g(a_n)]$, for all $a_1, \ldots, a_n \in M$. The equivalence can be proved by induction on φ.

Let $g : M \mapsto N$. For $X \subseteq M$, $Y \subseteq N$, we define $g^{\rightarrow}(X) = \{g(a) : a \in X\}$ (i.e., the image of X by g) and $g^{\leftarrow}(Y) = \{a \in M : g(a) \in Y\}$ (i.e., the co-image of Y by g).

Theorem 2 *Let $(M, \cdot_M, R_M)$ and $(N, \cdot_N, R_N)$ be relational phase spaces such that there exists an epimorphism of $(M, \cdot_M.R_M)$ onto $(N, \cdot_N, R_N)$. Then the first space is cyclic (resp. involutive) if and only if the second space is cyclic (resp. involutive). Furthermore $F(M, \cdot_M, R_M)$ is isomorphic to $F(N, \cdot_N, R_N)$.*

Proof Let g be an epimorphism of $(M, \cdot_M, R_M)$ onto $(N, \cdot_N, R_N)$. It is easy to show that $(M, \cdot_M, R_M)$ satisfies (RPS1) if and only if $(N, \cdot_N, R_N)$ satisfies (RPS1), and similarly for (RPS2). This yields the first part of the theorem for cyclic relational phase spaces. For involutive spaces we use Lemmas 2 and 3. We must show: $\phi_{R_M} = \psi_{R_M} \Leftrightarrow \phi_{R_N} = \psi_{R_N}$.

The conditions on the right-hand side of the equivalence in Lemma 2 can be expressed by formulas of the first-order language (without =), containing R as the only nonlogical symbol: $\varphi_1(u)$ expresses $\{u\}^{--\sim} \subseteq \{u\}^-$ (i.e., $\{u\}^-$ is ϕ_R–closed), $\varphi_2(u)$ expresses $\{u\}^{\sim\sim-} \subseteq \{u\}^\sim$ (i.e., $\{u\}^\sim$ is ψ_R–closed).

$\varphi_1(u)$: $\forall_x(\forall_y(\forall_z(R(z, u) \Rightarrow R(y, z)) \Rightarrow R(y, x)) \Rightarrow R(x, u))$
$\varphi_2(u)$: $\forall_x(\forall_y(\forall_z(R(u, z) \Rightarrow R(z, y)) \Rightarrow R(x, y)) \Rightarrow R(u, x))$

By Lemma 2, $\phi_{R_M} = \psi_{R_M}$ iff $\varphi_1[a]$ and $\varphi_2[a]$ are true in $(M, \cdot_M, R_M)$ for all $a \in M$. Since g is onto, $\phi_{R_N} = \psi_{R_N}$ iff $\varphi_1[g(a)]$ and $\varphi_2[g(a)]$ are true in $(N, \cdot_N, R_N)$ for all $a \in M$. The conditions on the right-hand side of these equivalences are equivalent, by Lemma 3.

We show that the map h defined by: $h(Y) = g^{\leftarrow}(Y)$ for $Y \in N_{R_N}$, is an isomorphism of $F(N.\cdot_N, R_N)$ and $F(M, \cdot_M, R_M)$.

We show: $g^{\leftarrow}(Y^{\sim}) = (g^{\leftarrow}(Y))^{\sim}$ for any $Y \subseteq N$. Let $z \in g^{\leftarrow}(Y^{\sim})$. Then $g(z) \in Y^{\sim}$, hence $R_N(y, g(z))$ for all $y \in Y$. This yields $R_N(g(u), g(z))$ for all $u \in g^{\leftarrow}(Y)$, hence $R_M(u, z)$ for all $u \in g^{\leftarrow}(Y)$. So $z \in (g^{\leftarrow}(Y))^{\sim}$. Let $z \in (g^{\leftarrow}(Y))^{\sim}$. Then $R_M(u, z)$ for all $u \in g^{\leftarrow}(Y)$, hence $R_N(g(u), g(z))$ for all $u \in g^{\leftarrow}(Y)$. Since $g^{\rightarrow}(g^{\leftarrow}(Y)) = Y$, $R_N(y, g(z))$ for all $y \in Y$. This yields $g(z) \in Y^{\sim}$, and consequently $z \in g^{\leftarrow}(Y^{\sim})$.

As a consequence, $h(X)$ is a fact for every fact X in $(N, \cdot_N, R_N)$. From $g^{\leftarrow}(X_1) = g^{\leftarrow}(X_2)$ we obtain $X_1 = X_2$, using $g^{\rightarrow}(g^{\leftarrow}(X_i)) = X_i$. So h is an injection.

We show $x \in Y^{\sim} \Leftrightarrow g(x) \in (g^{\rightarrow}(Y))^{\sim}$, for all $x \in M$, $Y \subseteq M$. We have: $x \in Y^{\sim}$ iff $R_M(y, x)$ for all $y \in Y$ iff $R_N(g(y), g(x))$ for all $y \in Y$ iff $R_N(u, g(x))$ for all $u \in g^{\rightarrow}(Y)$ iff $g(x) \in (g^{\rightarrow}(Y))^{\sim}$. So $Y^{\sim} = g^{\leftarrow}((g^{\rightarrow}(Y))^{\sim})$ for any $Y \subseteq M$. Accordingly h is a surjection.

We have shown $g^{\leftarrow}(Y^{\sim}) = (g^{\leftarrow}(Y))^{\sim}$, and $g^{\leftarrow}(Y^{-}) = (g^{\leftarrow}(Y))^{-}$ can be shown in a similar way. So h preserves negations. The equations:

$$g^{\leftarrow}(X\backslash Y) = g^{\leftarrow}(X)\backslash g^{\leftarrow}(Y) \quad g^{\leftarrow}(X/Y) = g^{\leftarrow}(X)/g^{\leftarrow}(Y) \tag{7}$$

are true for all $X, Y \subseteq N$. We only need the first equation. For $z \in M$, $z \in g^{\leftarrow}(X\backslash Y)$ iff $g(z) \in X\backslash Y$ iff $x \cdot_N g(z) \in Y$ for all $x \in X$ iff $g(u) \cdot_N g(z) \in Y$ for all $u \in g^{\leftarrow}(X)$ iff $g(u \cdot_M z) \in Y$ for all $u \in g^{\leftarrow}(X)$ iff $u \cdot_M z \in g^{\leftarrow}(Y)$ for all $u \in g^{\leftarrow}(X)$ iff $z \in g^{\leftarrow}(X)\backslash g^{\leftarrow}(Y)$. So h preserves $\backslash$. Clearly h preserves $\cap$. Therefore h is an isomorphism of the two algebras of facts. It preserves the least and the greatest element as well as the unit for multiplication in $F(N, \cdot_N, R_N)$ (if they exist), since every isomorphism preserves all first-order definable properties (in logic with =).

By $Fr[X]$ we denote the free groupoid generated by X. Its elements can be represented as variable-free terms of the first-order language having $\cdot$ as the only function symbol and all elements of X as individual constants. Clearly $\cdot$ is the multiplication in $Fr[X]$.

Theorem 3 *Every involutive (resp. cyclic) r.l. groupoid can be embedded in the algebra of facts of some involutive (resp. cyclic) phase space, based on a free groupoid.*

Proof Let **G** be an involutive r.l. groupoid. By Theorem 1, there exists an embedding h of **G** in $F(G, \odot, R_{\mathbf{G}})$. We consider the free groupoid $Fr[G]$. The map $g : Fr[G] \mapsto G$, defined by $g(a) = a$, for $a \in G$, $g(x \cdot y) = g(x) \odot g(y)$, for $x, y \in Fr[G]$, is an epimorphism of $(Fr[G], \cdot)$ onto $(G, \odot)$. We define a relation $S \subseteq Fr(G)^2$.

$$S(x, y) \Leftrightarrow R_{\mathbf{G}}(g(x), g(y)) \text{ for } x, y \in Fr[G] \tag{8}$$

Clearly g is an epimorphism of $(Fr[G], \cdot, S)$ onto $(G, \odot, R_{\mathbf{G}})$. By Theorem 2, the first space is involutive, and there exists an isomorphism h_1 of $F(G, \odot, R_{\mathbf{G}})$ onto $F(Fr[G], \cdot, S)$. So $h_1 \circ h$ is an embedding of $\mathbf{G}$ in $F(Fr[G], \cdot, S)$. $Fr[G]$ is a free groupoid. It follows that $F(Fr[G], \cdot, S) = F(Fr[G], \cdot, O)$, where $O = \{x \cdot y : S(x, y)\}$.

Let $\mathbf{G}$ be a cyclic r.l. groupoid. Then $(G, \odot, R_{\mathbf{G}})$ is cyclic. So $(Fr[G], \cdot, S)$ is cyclic, by Theorem 2. Consequently $(Fr[G], \cdot, O)$ is cyclic.

The embedding preserves the greatest and the least element as well as the multiplicative unit (if they exist in $\mathbf{G}$).

$(Fr[G], \cdot, O)$ is non-associative, non-commutative and not unital. If $\mathbf{G}$ is associative (resp. commutative, unital), then $F(Fr[G], \cdot, O)$ is associative (resp. commutative, unital), since it is isomorphic to $F(G, \odot, R_{\mathbf{G}})$.

Corollary 2 *The involutive (resp. cyclic) r.l. groupoids coincide with the subalgebras of algebras of facts in involutive (resp. cyclic) phase spaces, even based on free groupoids. The complete algebras of this kind coincide with the algebras of facts in the corresponding phase spaces.*

Proof The first part is obvious. If $\mathbf{G}$ is complete, then h is an isomorphism, by Theorem 1. Consequently, $h_1 \circ h$ is an isomorphism.

Let G be an involutive (resp. cyclic) r.l. monoid. In the proof of Theorem 3, we can replace $Fr[G]$ with G^+, i.e., the free semigroup generated by G. The elements of G^+ are all nonempty, finite sequences of elements of G, and $\cdot$ is the concatenation of sequences. Consequently, every involutive (resp. cyclic) r.l. monoid can be embedded into the algebra of facts in some involutive (resp. cyclic) phase space, based on a free semigroup. Since G is unital, then $a \odot b = c \odot d$ implies the equivalence of $R_{\mathbf{G}}(a, b)$ and $R_{\mathbf{G}}(c, d)$ (both are equivalent to $a \odot b \leq 0$). Consequently, $x \cdot y = u \cdot v$ in G^+ implies $S(x, y) \Leftrightarrow S(u, v)$, hence for $O = \{x \cdot y : S(x, y)\}$ we have $R_O = S$.

If G is not unital, not everything in the modified proof works. One shows that G can be embedded in $F(G^+, \cdot, S)$, which is a relational phase space. The definition $O = \{x \cdot y : S(x, y)\}$ does not imply $R_O = S$; we only have $S \subseteq R_O$, hence $F(G^+, \cdot, S) = F(G^+, \cdot, O)$ need not be true.

4 A Linguistic Interpretation of Cyclic Negation

Clark [6] adjusted concept lattices, studied in lattice theory, for modelling the family of syntactic categories of a language. We define some basic notions.

Σ denotes a finite lexicon. Σ^* is the set of all finite strings on Σ. The empty string is denoted by ϵ. Every set $L \subseteq \Sigma^*$ is called *a language* on Σ. We use x, y, z, u, v as variables for strings. The concatenation of strings x, y is denoted by $x \cdot y$ or xy. Clearly $\cdot$ is associative, and $(\Sigma^*, \cdot, \epsilon)$ is the free monoid generated by Σ. Therefore

$\mathcal{P}(\Sigma^*)$ with operations $\otimes, \backslash, /, \cup, \cap$ and the element $\{\epsilon\}$ is a complete residuated lattice ($\otimes, \backslash, /$ are defined by (2), (3)).

By *a context* one means a pair (x, y) such that $x, y \in \Sigma^*$. So $\Sigma^* \times \Sigma^*$ is the set of all contexts. This set is denoted by C.

Let $L \subseteq \Sigma^*$. One defines a binary relation between strings and contexts: $R_L(x, (u, v)) \Leftrightarrow uxv \in L$. For any sets $X \subseteq \Sigma^*$ and $Y \subseteq C$ one defines two sets.

$$X^{\rhd} = \{(u, v) \in C : \forall_{x \in X} R_L(x.(u, v))\}$$

$$Y^{\lhd} = \{x \in \Sigma^* : \forall_{(u,v) \in Y} R_L(x, (u, v))\}.$$

The operations $^{\rhd}, {}^{\lhd}$ form a Galois correspondence: $X \subseteq Y^{\lhd} \Leftrightarrow Y \subseteq X^{\rhd}$. So the operation $\eta(X) = X^{\rhd\lhd}$ is a closure operation on $\mathcal{P}(\Sigma^*)$ and the operation $\delta(Y) = Y^{\lhd\rhd}$ is a closure operation of $\mathcal{P}(C)$. We show that η is a nucleus. Let X be $\eta-$closed and $Y \subseteq \Sigma^*$. We show that $Y \backslash X$ and X/Y are $\eta-$closed. Like in Sect. 3 one shows that the $\eta-$closed sets are precisely the sets $Z^{\lhd}$ for $Z \subseteq C$. So $X = Z^{\lhd}$ for some $Z \subseteq C$. We define $Z_1 = \{(uy, v) : (u, v) \in Z, y \in Y\}$, $Z_2 = \{(u, yv) : (u, v) \in Z, y \in Y\}$. Clearly $Y \backslash Z^{\lhd} = Z_1^{\lhd}$ and $Z^{\lhd}/Y = Z_2^{\lhd}$. So $Y \backslash X$ and X/Y are $\eta-$closed. By (C4'), η is a nucleus.

The operations $\odot_\eta, \cup_\eta$ are defined as in Sect. 3. M_η denotes the family of $\eta-$closed subsets of Σ^*. Then M_η with operations $\odot_\eta, \backslash, /, \cup_\eta, \cap$ and the unit $\eta(\{\epsilon\})$ is a complete residuated lattice. Clark [6] also considers another residuated lattice whose elements are subsets of C, but it is not needed here. The pair of the two residuated lattices is *a syntactic concept lattice*.

According to Clark, the $\eta-$closed subsets of Σ^* are precisely the syntactic categories determined by the language L, interpreted as the set of sentences (resp. formulas) of a natural (resp. formal) language. This modifies the notion of a syntactic category as an equivalence class of the relation: $x \equiv y$ iff for all $u, v \in \Sigma^*$, $uxv \in L \Leftrightarrow uyv \in L$, used in mathematical linguistics, e.g., the theory of finite-state automata. Clark [6] focuses on some learning procedures for context-free grammars, applying syntactic concept lattices. Wurm [17] proves that Full Lambek Calculus is complete with respect to syntactic concept lattices.

We show that $Y^{\lhd}$ for any nonempty $Y \subseteq C$ equals $Y^{\sim}$ in an appropriately constructed cyclic associative unital phase space. We define a new lexicon $\Gamma = \Sigma \cup C$. So contexts are treated as lexical units.We consider the associative and unital phase space $(\Gamma^*, \cdot, O)$, where O consists of all strings $x(u, v)y$ such that $uyxv \in L$. This phase space satisfies (PS1), since it is associative. Clearly, it also satisfies (PS2). Therefore this phase space is cyclic. So the algebra of facts in this phase space is a complete cyclic r.l. monoid. Let $Y \subseteq C$ be nonempty. Then $Y \subseteq \Gamma^*$, since $C \subseteq \Gamma$. We have $Y^{\sim} = Y \backslash O$ and for any $z \in \Gamma^*$: $z \in Y \backslash O$ iff $(u, v)z \in O$, for all $(u, v) \in Y$, iff $z \in \Sigma^*$ and $uzv \in L$, for all $(u, v) \in Y$, iff $z \in Y^{\lhd}$. We have shown $Y^{\sim} = Y^{\lhd}$. If $Y = \emptyset$, then $Y^{\sim} = \Gamma^*$, but $Y^{\lhd} = \Sigma^*$.

A similar construction can be applied to non-associative syntactic concept lattices, not considered by Clark. Instead of Σ^* one takes the free unital groupoid generated by Σ; its subsets are called non-associative languages on Σ. A cyclic non-associative phase space is constructed, using some combinatorial properties of free groupoids. Non-associative languages can be interpreted as tree languages or languages, consisting of phrase structures. A context is an element of the free groupoid generated by $\Sigma \cup \{\_\}$, where _ is a special atom (a place for substitution), containing exactly one occurrence of _. We construct a cyclic non-associative phase space (G, O), where G is the free unital groupoid, generated by $\Sigma \cup C$, where C is the set of all contexts, and O is the smallest subset of G, containing all elements $c[\_] \cdot x$ such that $c[x] \in L$, satisfying (PS1) and (PS2). Using (PS1) and (PS2), every element of G, containing exactly one occurrence of a context, namely $c[\_]$, can be transformed to a unique element $c[\_] \cdot y$, where y contains no context; see Prop. 3 in [4].

In this short paper, we cannot discuss in detail the linguistic adequacy of Clark's ideas. If L consists of all grammatically correct sentences, then one easily finds some drawbacks, already observed in [6]. For instance, the context (John is a, student) seems to determine the category of adjective phrases, e.g., 'good', 'very good', but it is not true. One obtains the correct sentence 'John is a teacher and Mary is a student' by inserting 'teacher and Mary is a' into this context. The non-associative framework works better. The context (John (is (a (_ student)))) determines adjective phrases. For the associative framework, it is reasonable to take for L some large (finite) set of correct sentences, not containing all possible grammatical constructions.

5 Conclusions and Further Work

This paper contains some basic results on (relational) phase spaces and algebras of different linear logics. The main results show that every involutive (resp. cyclic) algebra can be embedded in the algebra of facts in a relational phase space and the algebra of facts in a phase space based on a free groupoid. The target algebra of facts preserves several properties of the source algebra: associativity, commutativity, multiplicative and additive units. Complete algebras coincide with algebras of facts (up to isomorphism). Girard [9] and his followers in the community of linear logics did not refer to algebras explicitly: they treated phase spaces as algebraic models, which interpret logical formulas as facts. Since models of PLL must be complete (to interpret exponentials), then the phase space models yield, actually, all algebraic models of this logic as algebras of facts. Non-associative (relational) phase spaces correspond to non-associative logics. Furthermore, they are essential in our Theorem 3: one cannot replace the free groupoid by a free semigroup, in general.

The proof of Theorem 2 applies some model-theoretic tools. This is quite natural for (relational) phase spaces. The author believes that other interesting results can be obtained in this way.

An interesting topic, omitted in this paper, are congruences on phase spaces. It follows from Theorem 2 that a quotient phase space yields the same algebra of facts (up to isomorphism) as the source phase space. A more thorough study of congruences on (relational) phase spaces is left for further research.

The phase space constructed in Sect. 4 from a syntactic concept lattice and its non-associative version are only one example of application of phase spaces in language theory. It would be interesting to find other examples, also for involutive (non-cyclic) phase spaces.

References

1. Abrusci VM (1991) Phase semantics and sequent calculus for pure noncommutative classical linear logic. J Symbolic Logic 56:1403–1451. https://doi.org/10.2307/2275485
2. Blaisdell E, Kanovich M, Kuznetsov SL, Pimentel E, Scedrov A (2022) Non-commutative, non-associative multi-modal linear logic. Lect Notes Comput Sci 13385:449–466. https://doi.org/10.1007/978-3-031-10769-6_27
3. Buszkowski W, Mathematical linguistics and proof theory. 16:683–736. https://doi.org/10.1016/8978-044481714-3/50016-3
4. Buszkowski W (2016) On classical nonassociative Lambek calculus. In: Amblard M, de Groote P, Pogodalla S, Retoré C (eds), Logical aspects of computational linguistics (LACL 2016), Lecture Notes in Computer Science, vol 10054. Springer, pp 68–84. https://doi.org/10.1007/978-3-662-53826-5_5
5. Buszkowski W (2019) On involutive nonassociative Lambek calculus. J Logic, Lang Inf 28:157–181. https://doi.org/10.1007/s10849-018-09278-9
6. Clark A (2011) A learnable representation for syntax using residuated lattices. In: de Groote P, Egg M, Kallmeyer L (eds) Formal grammar. Lecture Notes in Computer Science, vol 5591. Springer, pp 183–198. https://doi.org/10.1007/978-3-642-20169-1_12
7. Galatos N, Jipsen P (2013) Residuated frames with applications to decidability. Trans Am Math Soc 365:1219–1249. https://doi.org/10.1090/S0002-9947-2012-05573-5
8. Galatos N, Jipsen P, Kowalski T, Ono H (2007) Residuated lattices: an algebraic glimpse at substructural logics. In: Studies in Logic and The Foundations of Mathematics, vol 151. Elsevier. ISBN:978-0-444-52141-5
9. Girard J-Y (1987) Linear logic. Theor Comput Sci 50:1–102. https://doi.org/10.1016/0304-3975(87)90045-4
10. Lafont Y (1997) The finite model property for various fragments of linear logic. J Symbolic Logic 62:1202–1208. https://doi.org/10.2307/2275637
11. Lambek J (1958) The mathematics of sentence structure. Am Math Mon 65:154–170
12. Lambek J (1961) On the calculus of syntactic types. In: Jakobson R (ed) Structure of language and its mathematical aspects, proceedings of symposium in applied mathematics, American Mathematical Society, pp 166–178. https://doi.org/10.1090/psapm/012
13. Moortgat M, Categorial type logic. 16:93–177. https://doi.org/10.1016/8978-044481714-3/50005-9
14. Płaczek P (2021) Extensions of Lambek calculi: sequent systems, conservativeness and computational complexity. Ph.D. Thesis, Adam Mickiewicz University, Poznań
15. van Benthem J (1986) Essays in logical semantics. D. Reidel. https://doi.org/10.1007/978-94-009-4540-1
16. van Benthem J, ter Meulen A (eds) (1997) Handbook of logic and language. Elsevier, The MIT Press. ISBN: 9780444817143
17. Wurm C (2017) Language-theoretic and finite relation models for the (full) Lambek calculus. J Logic, Lang Inf 26:179–214. https://doi.org/10.1007/s10849-017-8249-z
18. Yetter DN (1990) Quantales and (non-commutative) linear logic. J Symbolic Logic 55:41–64. https://doi.org/10.2307/2274953

Applying Quantales to Lambek Calculi with Cyclic Negation

Paweł Płaczek

Abstract The paper investigates the application of quantale semantics to non-classical substructural logics, specifically Lambek calculi extended with cyclic negation. We introduce models based on involutive quantales and explore their correspondence with cyclic linear logic frameworks. The approach provides a unified perspective on structural rules and dualities in categorial grammars. Particular attention is given to the interpretation of non-associativity and non-commutativity in linguistic and proof-theoretic contexts.

Keywords Quantales · Lambek calculus · Cyclic negation · Substructural logic · Involutive residuation · Categorial grammar · Non-associativity

1 Introduction

Lambek Calculus L was introduced by Lambek [9] under the name *Syntactic Calculus*. L is a propositional logic with three connectives $\otimes$ (product), $\multimap$ and (residuations of product). Lambek's symbols for residuations are $\backslash$ and $/$, and this notation is used in the literature on substructural logics; see, e.g., Galatos et al. [7].

Product is also called tensor, multiplicative conjunction or intensional conjunction. The corresponding operation in algebras satisfies none of the three laws of conjunction $\wedge$: $a \wedge b \leq a$, $a \wedge b \leq b$, $a \leq b$ and $a \leq c$ imply $a \leq b \wedge c$. One assumes the associative law $(a \otimes b) \otimes c = a \otimes (b \otimes c)$, but not the commutative law $a \otimes b = b \otimes a$. Its residuations can be understood as substructural implications. If product is commutative, then the two implications collapse: $a \multimap b = b \quad a$; one writes $a \rightarrow b$.

Lambek [10] introduced the nonassociative version of this logic, nowadays called Nonassociative Lambek Calculus (NL). In linguistics, NL is employed for parsing expressions given in the form of trees (phrase structures), while L corresponds to expressions as strings (of lexical items). From the perspective of logic, NL can be seen

P. Płaczek (✉)
WSB Merito University in Poznań, Poznań, Poland
e-mail: pawel.placzek@poznan.merito.pl

R. Loukanova (ed.), *Computational Linguistics, Information, Reasoning, and AI 2024*, Studies in Computational Intelligence 1280,
https://doi.org/10.1007/978-3-032-21143-9_4

as the pure logic of residuation and L as its stronger version for associative product. In [10] Lambek provided a sequent system for NL and proved cut elimination.

L and NL were mainly studied by scholars interested in categorial grammars; their mathematical theory is highly nontrivial and reveals an interplay of logic and formal linguistics. Extensions of Lambek calculi were extensively studied in logic.

In Galatos et al. [7] Full Lambek Calculus FL is defined as the extension of L, admitting the constant 1 (interpreted as the unit for product) and lattice connectives $\vee$ and $\wedge$. Optionally one adds the constant 0 and defines substructural negations $\sim A = A \multimap 0$ and $-A = 0 \quad A$. Substructural logics are defined as extensions of FL by new axioms and rules. Their algebraic models are different classes of residuated lattices. Assuming commutativity of $\otimes$, these two negations collapse.

Substructural negations need not satisfy the double negation law(s); in algebras $\sim - a = a$, $-\sim a = a$. Assuming these laws leads to *linear logics*. Propositional linear logic PLL of Girard [8] can be presented as an extension of the commutative FL with 0 and exponentials !, ? by the double negation law. Exponentials are some unary modalities, which make it possible to interpret classical propositional logic in PLL. The fragment without exponentials is called Multiplicative-Additive Linear Logic MALL. (In the literature on linear logics, Lambek connectives and constants 1, 0 are said to be multiplicative, while lattice connectives and constants $\top$, $\bot$ additive; $\top$ and $\bot$ are interpreted as the greatest and the least element, respectively.) Cyclic MALL (CyMALL) of Yetter [14] drops commutativity but assumes $\sim a = -a$ in algebras, and Noncommutative MALL of Abrusci [1] omits this assumption.

In this paper we consider nonassociative variant of Cyclic MALL (NCyMALL). Buszkowski [5] proves that the multiplicative, constant–free fragment of this logic (CyNL) is a strongly conservative extension of NL. The extension is said to be *strongly conservative*, if in the extension we do not obtain new consequences of the assumptions in the language of the original logic. In the present paper we extend the result from [5] and we prove that NCyMALL is a strongly conservative extension of Full Nonassociative Lambek Calculus (FNL). This result holds also for associative CyMALL.

These results cannot be extended for additive constants: $\top$ and $\bot$. Abrusci [2] has previously demonstrated that CyMALL is not a conservative extension of FL when considering the inclusion of additive constants. However, when the additive constants are omitted, CyMALL exhibits conservativeness. Abrusci's work presents a sequent that is provable in CyMALL but not in FL with additive constants. It follows that strong conservativeness does not hold for nonassociative versions.

Buszkowski [3] proves that Lambek Calculus has undecidable consequence relation. As a consequence, all strongly conservative extensions of this logics also have undecidable consequence relation. So, as the consequence of the main result of this paper, we obtain that CyMALL and its fragments have undecidable consequence relations. Chvalovský [6] proves the same result for Full Nonassociative Lambek Calculus, so NCyMALL and its fragments with additive connectives have undecidable consequence relations. The positive result in this matter has been proved in Buszkowski [4]—the finitary consequence relation of NL is decidable in PTIME. The similar result holds for CyNL; see [5].

Noncommutative MALL is a weaker logic than Cyclic MALL, so Noncommutative MALL, its nonassociative variant and their fragments are also strongly conservative extensions of the respective variants of Lambek Calculus.

The results of this paper have been already proved using syntactic methods in [13]. This paper aims to provide an extended meaning for quantales and a practical example of application, hence we provide a simpler proof using algebraic methods.

2 Logics

2.1 *Full Nonassociative Lambek Calculus*

The atomic FNL-formulas are variables and the constant: 1. The connectives are $\otimes, \multimap, \;\; , \vee, \wedge$. The connectives $\otimes, \multimap, \;\;$ are said to be *multiplicative* and $\vee, \wedge$ are said to be *additive*. The multiplicative constant is 1.

An *FNL-bunch* is an element of a free unital groupoid generated from the set of all FNL-formulas. The unit of this free groupoid is called the *empty bunch* and denoted ϵ. A *context* is a bunch with one occurrence of a special formula $-$. By $\Gamma[\Delta]$ we denote the context $\Gamma[-]$ with Δ substituted in the place of $-$. Let (Γ, A) be a pair of a bunch and a formula. We write $\Gamma \Rightarrow A$ and call it an *FNL-sequent*. We reserve capital Latin letters for formulas and capital Greek letters for bunches. In the sequent $\Gamma \Rightarrow A$, the bunch Γ is the *antecedent* and the formula A is the *consequent*. Since FNL does not assume associativity or commutativity of the implication product, the structure of bunches reflects the parse tree of linguistic expressions, which is essential for modeling linguistic phenomena in categorial grammars.

The axioms and the rules of FNL are as follows, in (id), …, $(\Rightarrow \wedge)$.

$$(\text{id})\quad A \Rightarrow A \qquad\qquad (\text{cut})\quad \frac{\Gamma \Rightarrow A \quad \Delta[A] \Rightarrow C}{\Delta[\Gamma] \Rightarrow C}$$

$$(\otimes\Rightarrow)\quad \frac{\Gamma[(A, B)] \Rightarrow C}{\Gamma[A \otimes B] \Rightarrow C} \qquad\qquad (\Rightarrow\otimes)\quad \frac{\Gamma \Rightarrow A \quad \Delta \Rightarrow B}{\Gamma, \Delta \Rightarrow A \otimes B}$$

$$(\multimap\Rightarrow)\quad \frac{\Gamma[B] \Rightarrow C \quad \Theta \Rightarrow A}{\Gamma[(\Theta, A \multimap B)] \Rightarrow C} \qquad\qquad (\Rightarrow\multimap)\quad \frac{A, \Gamma \Rightarrow B}{\Gamma \Rightarrow A \multimap B}$$

$$(\;\;\Rightarrow)\quad \frac{\Gamma[A] \Rightarrow C \quad \Theta \Rightarrow B}{\Gamma[(A \;\;\; B, \Theta)] \Rightarrow C} \qquad\qquad (\Rightarrow\;\;)\quad \frac{\Gamma, B \Rightarrow A}{\Gamma \Rightarrow A \;\;\; B}$$

$$(1\Rightarrow)\quad \frac{\Gamma[\Delta] \Rightarrow C}{\Gamma[(1, \Delta)] \Rightarrow C} \quad \frac{\Gamma[\Delta] \Rightarrow C}{\Gamma[(\Delta, 1)] \Rightarrow C} \qquad\qquad (\Rightarrow 1)\quad \epsilon \Rightarrow 1$$

$$(\vee\Rightarrow)\quad \frac{\Gamma[A] \Rightarrow C \quad \Gamma[B] \Rightarrow C}{\Gamma[A \vee B] \Rightarrow C} \qquad\qquad (\Rightarrow\vee)\quad \frac{\Gamma \Rightarrow A}{\Gamma \Rightarrow A \vee B} \quad \frac{\Gamma \Rightarrow B}{\Gamma \Rightarrow A \vee B}$$

$$(\wedge\Rightarrow)\quad \frac{\Gamma[A] \Rightarrow C}{\Gamma[A \wedge B] \Rightarrow C} \quad \frac{\Gamma[B] \Rightarrow C}{\Gamma[A \wedge B] \Rightarrow C} \qquad\qquad (\Rightarrow\wedge)\quad \frac{\Gamma \Rightarrow A \quad \Gamma \Rightarrow B}{\Gamma \Rightarrow A \wedge B}$$

The models of FNL are residuated lattices, which provide the algebraic semantics for the logic. In such structures, the residuated properties ensure that every provable sequent is valid in all models, and conversely, every valid consequence in the semantics can be derived syntactically from the axioms and rules.

Definition 2.1 Let $\mathbf{L} = (L, \otimes, \multimap, \quad, \vee, \wedge, 1, \leq)$ be a structure such that $(L, \vee, \wedge)$ is a lattice, $(L, \leq)$ is a lattice order, $(L, \otimes, 1)$ is a unital groupoid and the following property holds:

$$\text{(RES)} \qquad a \otimes b \leq c \text{ iff } b \leq a \multimap c \text{ iff } a \leq c \quad b.$$

The structure $\mathbf{L}$ is a *residuated lattice*.

The property (RES) is called *the residuation law*, and it establishes the fundamental relationship between the three operations: the tensor product $\otimes$ and its two residuals $\multimap$ and .

The following properties of residuated lattices will be useful:

$$a \otimes (a \multimap b) \leq b \qquad\qquad (b \quad a) \otimes a \leq b \qquad (2.1)$$
$$b \leq a \multimap (a \otimes b) \qquad\qquad b \leq (b \otimes a) \quad a \qquad (2.2)$$
$$a \leq (b \quad a) \multimap b \qquad\qquad a \leq b \quad (a \multimap b) \qquad (2.3)$$
$$a \multimap b = (b \quad (a \multimap b)) \multimap b \qquad b \quad a = b \quad ((b \quad a) \multimap b) \qquad (2.4)$$

for all $a, b \in L$.

Let μ be a homomorphism from the free algebra of formulas into the residuated lattice $\mathbf{L}$. We extend μ to bunches by $\mu((\Gamma, \Delta)) = \mu(\Gamma) \otimes \mu(\Delta)$ and $\mu(\epsilon) = 1$. We say μ is a *valuation*.

Let $\Gamma \Rightarrow A$ be an FNL-sequent. We say $\Gamma \Rightarrow A$ is *satisfied in* $\mathbf{L}$ *under the valuation* μ, if $\mu(\Gamma) \leq \mu(A)$. We say $\Gamma \Rightarrow A$ is *valid in* $\mathbf{L}$, if it is satisfied under every valuation μ.

Let Φ be a set of FNL-sequents. We say an FNL-sequent $\Gamma \Rightarrow A$ is a *consequence* of Φ, if, and only if, $\Gamma \Rightarrow A$ is provable from Φ. We write $\Phi \vdash_{\mathrm{FNL}} \Gamma \Rightarrow A$. We say that Φ *entails* $\Gamma \Rightarrow A$, if for all residuated lattices and all valuations μ the sequent $\Gamma \Rightarrow A$ is satisfied under μ whenever all sequents from Φ are satisfied under μ. We write $\Phi \models_{\mathrm{FNL}} \Gamma \Rightarrow A$.

FNL is strongly complete with respect to residuated lattices. The proof of this fact is a standard construction of Lindenbaum-Tarski algebra.

We prove soundness. Let $\mathbf{L}$ be a residuated lattice and let μ be a valuation. (id) is satisfied, since $\mu(A) \leq \mu(A)$ for all formulas A. We check the rules. The cut rule (cut) is valid because of the monotonicity of $\otimes$. $(\otimes \Rightarrow)$ is valid by definition of μ, $(\Rightarrow \otimes)$ follows from the monotonicity of $\otimes$, the rules for $\multimap$ and   are valid because of (RES). Since $\mu(1)$ is the unit for $\otimes$, then $\mu((1, \Gamma)) = \mu(\Gamma)$, so $(1 \Rightarrow)$ is valid; $(\Rightarrow 1)$ follows from the definition of μ and reflexivity of the order. Rules for $\wedge$ and $\vee$ are valid by the properties of residuated lattices. Hence, by induction on proofs, one shows $\Gamma \Rightarrow A$ is satisfied under every valuation μ, since every axiom is satisfied and every rule is valid.

Now we construct the Lindenbaum-Tarski algebra. One notices that ($\otimes \Rightarrow$) is reversible, so we can represent every sequent $\Gamma \Rightarrow A$ as $G \Rightarrow A$, where G arises from Γ by using ($\otimes \Rightarrow$) as long as we do not obtain a formula. We define $A \leq B$ iff $\Phi \vdash_{\mathrm{FNL}} A \Rightarrow B$. Also, we define $A \equiv B$ iff $A \leq B$ and $B \leq A$. One shows $\equiv$ is a congruence in the algebra of formulas. We take the quotient algebra with the quotient ordering and obtain residuated lattice. We define $\mu(p) = [p]_{\equiv}$. By induction on formulas, one proves $\mu(A) = [A]_{\equiv}$. Hence, if $\Phi \vdash_{\mathrm{FNL}} G \Rightarrow A$, then $[G]_{\equiv} \leq [A]_{\equiv}$. If $\Phi \nvdash_{\mathrm{FNL}} G \Rightarrow A$, then $[G]_{\equiv} \nleq [A]_{\equiv}$.

2.2 *Nonassociative Cyclic MALL*

Let $\mathcal{V}$ be an arbitrary, countable set of variables. We define the set of atoms $\mathcal{V}'$ as follows: if $p \in \mathcal{V}$, then $p \in \mathcal{V}'$ and $p^{\sim} \in \mathcal{V}'$ and $1, 0 \in \mathcal{V}'$. The variables are called *positive* atoms and the negated variables are called *negative* atoms. We construct the set of NCyMALL-formulas from $\mathcal{V}'$ by the binary connectives: $\otimes$, , $\wedge$ and $\vee$.

One notices we do not consider negation as connective. The systems we introduce are in negation normal form, i.e., the negations are only in the form of atoms.

We define the metalanguage negation $^{\sim}$:

$$
\begin{array}{rcl}
(p)^{\sim} = p^{\sim} & \qquad & (p^{\sim})^{\sim} = p \\
1^{\sim} = 0 & & 0^{\sim} = 1 \\
(A \otimes B)^{\sim} = B^{\sim} \quad A^{\sim} & & (A \quad B)^{\sim} = B^{\sim} \otimes A^{\sim} \\
(A \wedge B)^{\sim} = A^{\sim} \vee B^{\sim} & & (A \vee B)^{\sim} = A^{\sim} \wedge B^{\sim}
\end{array}
$$

One shows $A^{\sim\sim} = A$ for all formulas A.

We define the NCyMALL-bunches analogously to FNL-bunches. An NCyMALL-sequent is any nonempty bunch. In the sequents, we often omit the outer parenthesis.

The axioms and rules for NCyMALL are as follows, in (a-id), ..., (r-$\wedge$).

(a-id) $\quad p, p^{\sim}$

(r-$\otimes$) $\quad \dfrac{\Gamma[(A, B)]}{\Gamma[A \otimes B]}$

(r- 1) $\quad \dfrac{\Gamma[B] \quad \Delta, A}{\Gamma[(\Delta, A \quad B)]}$

(r- 2) $\quad \dfrac{\Gamma[A] \quad B, \Delta}{\Gamma[(A \quad B, \Delta)]}$

(r-shift) $\quad \dfrac{(\Gamma, \Delta), \Theta}{\Gamma, (\Delta, \Theta)}$

(r-cyc) $\quad \dfrac{\Gamma, \Delta}{\Delta, \Gamma}$

(r-cut) $\quad \dfrac{\Gamma[A] \quad A^{\sim}, \Delta}{\Gamma[\Delta]}$

(a-0) $\quad 0$

(r-1) $\quad \dfrac{\Gamma[\Delta]}{\Gamma[(1, \Delta)]} \quad \dfrac{\Gamma[\Delta]}{\Gamma[(\Delta, 1)]}$

(r-$\wedge$) $\quad \dfrac{\Gamma[A]}{\Gamma[A \wedge B]} \quad \dfrac{\Gamma[A]}{\Gamma[B \wedge A]}$

(r-$\vee$) $\quad \dfrac{\Gamma[A] \quad \Gamma[B]}{\Gamma[A \vee B]}$

One shows that (r-cyc) is reversible.

The associative variant of NCyMALL—CyMALL is obtained by replacing bunches with finite sequences of formulas. The rules look similar.

The models of NCyMALL are cyclic residuated lattices.

Definition 2.2 Let $\mathbf{L} = (L, \otimes, \vee, \wedge, {}^{\sim}, 1, \leq)$ be a structure such that $(L, \vee, \wedge, \leq)$ is a lattice, $(L, \otimes, 1)$ is a unital groupoid and the following properties hold:

$$\text{(COM)} \qquad a \otimes b \leq c \text{ iff } c^{\sim} \otimes a \leq b^{\sim} \text{ iff } b \otimes c^{\sim} \leq a^{\sim}$$

$$\text{(CONT)} \qquad \text{if } a \leq b, \text{ then } b^{\sim} \leq a^{\sim}$$

$$\text{(DN)} \qquad a^{\sim\sim} = a$$

Then $\mathbf{L}$ is a *cyclic residuated lattice*.

The property (COM) is called *the compatibility law*. One defines $a \quad b = (b^{\sim} \otimes a^{\sim})^{\sim}$ and $0^{\sim} = 1$. The models of CyMALL are cyclic residuated lattices, where $(L, \otimes, 1)$ is a monoid.

One defines:

$$a \multimap b = a^{\sim} \quad b \qquad b \quad a = b \quad a^{\sim} \tag{2.5}$$

and proves (RES). Hence, every cyclic residuated lattice is a residuated lattice.

Let $(L, {}^{\sim}, \leq)$ be a structure such that $(L, \leq)$ is a poset and ${}^{\sim}$ satisfies (CONT) and (DN). We say ${}^{\sim}$ is a *De Morgan negation*.

Let μ be a homomorphism from the free algebra of NCyMALL-formulas into the cyclic residuated lattice $\mathbf{L}$. We extend μ to bunches by $\mu((\Gamma, \Delta)) = \mu(\Gamma) \otimes \mu(\Delta)$ and $\mu(\epsilon) = 1$. We say μ is a *valuation*.

Let Γ be an NCyMALL-sequent. We say Γ is *satisfied in* $\mathbf{L}$ *under the valuation* μ, if $\mu(\Gamma) \leq 0$. We say Γ is *valid in* $\mathbf{L}$, if it is satisfied under every valuation μ.

Let Φ be a set of NCyMALL-sequents. We say an NCyMALL-sequent Γ is a *consequence* of Φ, if, and only if, Γ is provable from Φ. We write $\Phi \vdash_{\mathrm{NCyMALL}} \Gamma$. We say that Φ *entails* Γ, if for all cyclic residuated lattices and all valuations μ the sequent Γ is satisfied under μ whenever all sequents from Φ are satisfied under μ. We write $\Phi \models_{\mathrm{NCyMALL}} \Gamma$.

We define the translation of the sequents of FNL into the sequents of NCyMALL. The translation of the formulas preserves all connectives with exception for $\multimap$ and . We translate $A \multimap B$ to $A^{\sim} \quad B$ and $A \quad B$ to $A \quad B^{\sim}$. A sequent $\Gamma \Rightarrow A$ is translated to $A^{\sim}, \Gamma$.

NCyMALL is strongly complete with respect to cyclic residuated lattices. The proof of this fact is a standard construction of Lindenbaum-Tarski algebra.

We show (a-id) is valid in cyclic residuated lattices. We have $\mu(p) \leq \mu(p)$ and $1 \otimes \mu(p) \leq \mu(p)$. By (COM), $\mu(p) \otimes \mu(p)^{\sim} \leq 0$. By definition of a valuation, $\mu(p) \otimes \mu(p^{\sim}) \leq 0$, so (a-id) is valid. (r-$\otimes$) is valid by the definition of a valuation. (r-1) is valid, since 1 is the unit for $\otimes$. Rules (r-$\wedge$) and (r-$\vee$) express the lattice order properties. One proves $a^{\sim} \otimes (a \quad b) \leq b$ and $(b \quad a) \otimes a^{\sim} \leq b$ by (2.1) and (2.5).

This proves the validity of (r-   1) and (r-   2). We show (r-shift). The following are equivalent:

$$(\mu(\Gamma) \otimes \mu(\Delta)) \otimes \mu(\Theta) \leq 0$$
$$0^{\sim} \otimes (\mu(\Gamma) \otimes \mu(\Delta)) \leq (\mu(\Theta))^{\sim}$$
$$1 \otimes (\mu(\Gamma) \otimes \mu(\Delta)) \leq (\mu(\Theta))^{\sim}$$
$$\mu(\Gamma) \otimes \mu(\Delta) \leq (\mu(\Theta))^{\sim}$$
$$\mu(\Delta) \otimes (\mu(\Theta))^{\sim\sim} \leq (\mu(\Gamma))^{\sim}$$
$$\mu(\Delta) \otimes \mu(\Theta) \leq (\mu(\Gamma))^{\sim}$$
$$(\mu(\Delta) \otimes \mu(\Theta)) \otimes 0^{\sim} \leq (\mu(\Gamma))^{\sim}$$
$$\mu(\Gamma) \otimes (\mu(\Delta) \otimes \mu(\Theta)) \leq 0$$

We construct a Lindenbaum-Tarski algebra. We define relation $A \leq B$ iff $\Phi \vdash A, B^{\sim}$. The rest of the construction is similar as for FNL.

3 Quantales

Yetter [14] used quantales[1] to describe models of CyMALL. We adapt this notion to fit NCyMALL.

3.1 Nonassociative Quantales

Definition 3.1 Let $(Q, \bigvee, \leq)$ be a complete lattice and let $\otimes$ be an operation on Q, which distributes over $\bigvee$ (on both sides). $(Q, \bigvee, \otimes, 1, \leq)$ is a *nonassociative quantale*, if $(Q, \otimes, 1)$ is a unital groupoid.

Remark 3.2 The name *nonassociative quantale* may be misleading. We do not require $\otimes$ to be nonassociative, but we just simply do not assume that this operation is associative. More intuitive name: *not necessarily associative quantale* is too long for practical use. The simple term *quantale* stands for the analogous structure where $\otimes$ is associative.

Let **Q** be a (nonassociative) quantale. For every $a, c \in Q$ we define:

$$a \multimap c = \bigvee\{b \in Q : a \otimes b \leq c\} \qquad c \quad a = \bigvee\{b \in Q : b \otimes a \leq c\}$$

We prove (RES) for (nonassociative) quantales. Let $a \otimes b \leq c$. Then, $b \in \{x \in Q : a \otimes x \leq c\}$. So, $b \leq a \multimap c$. Now let $b \leq a \multimap c$. Let $d \in \{x : a \otimes x \leq c\}$, then $\bigvee(a \otimes d) \leq c$. By distributivity of $\otimes$ over $\bigvee$ we obtain $a \otimes \bigvee d \leq c$. Hence,

[1] Quantales has been introduced by Mulvey [11], but Yetter [14] follows Niefield and Rosenthal [12].

$a \otimes (a \multimap c) \leq c$. By monotonicity, $a \otimes b \leq a \otimes (a \multimap c) \leq c$. The second part is proved analogously.

Therefore, every (nonassociative) quantale is a residuated lattice.

3.2 Special Elements of Nonassociative Quantales

Definition 3.3 Let $a \in Q$ be an element of the nonassociative quantale $\mathbf{Q}$. Then:

- a is *dualizing*, if for all $b \in Q$:

$$a \quad (b \multimap a) = b = (a \quad b) \multimap a$$

- a is *cyclic*, if for all $b, c \in Q$: if $b \otimes c \leq a$, then $c \otimes b \leq a$.
- a is *shifting*, if for all $b, c, d \in Q$: $(b \otimes c) \otimes d \leq a$ if, and only if, $b \otimes (c \otimes d) \leq a$.

Proposition 3.4 *The element a is cyclic, if, and only if,* $b \multimap a = a \quad b$ *for all b.*

Let $s \in Q$ be a shifting element. We prove the following properties for all $a, b \in Q$:

$$(a \otimes b) \multimap s = b \multimap (a \multimap s) \tag{3.1}$$

$$s \quad (a \otimes b) = (s \quad b) \quad a \tag{3.2}$$

We show (3.1). By (2.1):

$$(a \otimes b) \otimes ((a \otimes b) \multimap s) \leq s.$$

Since s is shifting:

$$a \otimes [b \otimes ((a \otimes b) \multimap s)] \leq s.$$

By (RES):

$$b \otimes ((a \otimes b) \multimap s) \leq a \multimap s$$

and:

$$(a \otimes b) \multimap s \leq b \multimap (a \multimap s).$$

Conversely, by (2.1):

$$b \otimes (b \multimap (a \multimap s)) \leq a \multimap s.$$

By (RES):

$$a \otimes (b \otimes (b \multimap (a \multimap s))) \leq s.$$

Since s is shifting:

$$(a \otimes b) \otimes (b \multimap (a \multimap s)) \leq s.$$

By (RES):

$$b \multimap (a \multimap s) \leq (a \otimes b) \multimap s.$$

(3.2) is proved analogously.

3.3 Cyclic Nonassociative Quantales

Yetter [14] uses the term *Girard's quantale* for quantales with De Morgan negation, where the product is associative. In this section we adjust this idea to nonassociative framework.

Definition 3.5 Let $(Q, \bigvee, \otimes, 1, \leq)$ be a nonassociative quantale and let $^{\sim}$ be a De Morgan negation on $(Q, \bigvee, \leq)$. We say $(Q, \bigvee, \otimes, ^{\sim}, 1, \leq)$ is a *cyclic nonassociative quantale*, if (COM) holds.

We define   and 0 as above. We prove that 0 is cyclic in nonassociative quantales. Let $a, b \in Q$ and $a \otimes b \leq 0$. By (COM), $b \otimes 1 \leq a^{\sim}$. Then, $1 \otimes b \leq a^{\sim}$. By (COM), $b \otimes a \leq 0$.

We provide a method for obtaining cyclic nonassociative quantales from nonassociative quantales.

Definition 3.6 Let **Q** be a nonassociative quantale and let j be a unary operation on Q. We say j is a quantic nucleus, if:

$$\begin{array}{ll} \text{(QN1)} & a \leq j(a) \\ \text{(QN2)} & \text{if } a \leq b, \text{ then } j(a) \leq j(b) \\ \text{(QN3)} & j(j(a)) \leq j(a) \\ \text{(QN4)} & j(a) \otimes j(b) \leq j(a \otimes b) \end{array}$$

for all $a, b \in Q$.

Let $(P, \leq)$ be a poset. The operation $j : P \multimap P$ is called a *closure operation*, if j satisfies (QN1)–(QN3) for all $a, b \in P$. An element $a \in P$ is said to be j-closed, if $a = j(a)$ (equivalently: $j(a) \leq a$). Clearly, the j-closed elements are precisely the elements $j(a)$ for $a \in P$. If $(P, \leq)$ is a complete lattice, then the family of j-closed elements is closed under $\bigwedge$.

Proposition 3.7 *Let $(Q, \bigvee, \otimes, 1, \leq)$ be a nonassociative quantale. Let j be a closure operation on $(Q, \leq)$. Then (QN4) is equivalent to:*

(QN4') if $j(a) \leq a$, then $j(b \multimap a) \leq b \multimap a$ and $j(a \quad b) \leq a \quad b$

for all $a, b \in Q$.

Proof Assume (QN4). Let $a, b \in Q$ and a be j-closed. We will show $j(b \multimap a) \leq b \multimap a$. We have

$$b \otimes j(b \multimap a) \le j(b) \otimes j(b \multimap a) \le j(b \otimes (b \multimap a)) \le j(a) = a.$$

Consequently, $j(b \multimap a) \le b \multimap a$, by (RES). $j(a \quad b) \le a \quad b$ can be shown in a similar way.

Assume (QN4'). Let $a, b \in Q$. From $a \otimes b \le j(a \otimes b)$ we get $b \le a \multimap j(a \otimes b)$. Hence, $j(b) \le a \multimap j(a \otimes b)$, by (QN2) and (QN4'). So, $a \otimes j(b) \le j(a \otimes b)$ by (RES). This yields $a \le j(a \otimes b) \quad j(b)$, which leads to $j(a) \otimes j(b) \le j(a \otimes b)$, using (QN4') and (RES) as above. □

Proposition 3.8 *Let* $\mathbf{Q}$ *be a nonassociative quantale and let* $f \in Q$ *be a cyclic, shifting element. We define* $j(x) = f \quad (x \multimap f)$ *for all* $x \in Q$. *Then* j *is a quantic nucleus and the lattice of* j*-closed elements of* Q *forms a cyclic quantale.*[2]

Proof Let $\mathbf{Q} = (Q, \bigvee, \cdot, 1, \le)$ be a nonassociative quantale and let $f \in Q$ be a cyclic, shifting element. We prove j is a quantic nucleus.

(QN1). follows from (2.3).
(QN2). Let $a, b \in Q$ and $a \le b$. Then $b \multimap f \le a \multimap f$ and:

$$j(a) = f \quad (a \multimap f) \le f \quad (b \multimap f) = j(b).$$

(QN3). Let $a \in Q$. By (2.3) we have:

$$a \multimap f \le (f \quad (a \multimap f)) \multimap f$$

and then $j(j(a)) \le j(a)$, since is antitone.
(QN4). First, we observe that for $a \in Q$ the following conditions are equivalent:

(i) a is j-closed,
(ii) $a = x \multimap f$ for some $x \in Q$,
(iii) $a = f \quad y$ for some $y \in Q$.

Clearly, (i) implies (ii). We show the converse implication. Assume $a = x \multimap f$. Then $j(a) = (f \quad (x \multimap f)) \multimap f) = x \multimap f = a$, by (2.4). The equivalence of (i) and (iii) can be shown in a similar way. It suffices to prove (QN4'). Assume that a is j-closed and $b \in Q$. We show that $b \multimap a$ is j-closed. There exists $x \in Q$ such that $a = x \multimap f$. So, $b \multimap a = b \multimap (x \multimap f) = (x \otimes b) \multimap f$, by (3.1). Hence, $b \multimap a$ is j-closed. For $a \quad b$ the argument is similar.

We prove f is j-closed. We consider $1 \multimap f = \bigvee\{a \in Q : a \cdot 1 \le f\} = f$, so f is j-closed.

We define $\mathbf{Q}_j = (Q_j, \bigvee_j, \otimes, {}^{\sim}, I, \le)$, where Q_j is a set of j-closed elements of Q, $\bigvee_f A = j(\bigvee A)$ for all $A \subseteq Q_j$, $a \otimes b = j(a \cdot b)$ for all $a, b \in Q_j$, $I = j(1)$ and $a^{\sim} = a \multimap f$ for all $a \in Q_j$.

We show $\otimes$ distributes over $\bigvee_j$. Let $a, b_\alpha \in Q$ for $\alpha \in \mathcal{A}$.

---

[2] This proposition is a modification of Proposition 1.7 from Yetter [14]. The original result is for associative variant of quantales.

$$a \otimes (\bigvee_j b_\alpha) = j(a \cdot j(\bigvee b_\alpha)) = j(a \cdot \bigvee b_\alpha) = j(\bigvee (a \cdot b_\alpha)) \leq$$
$$\leq j(\bigvee j(a \cdot b_\alpha)) = \bigvee_j (a \otimes b_\alpha)$$

The argument for the other side is similar. We show (COM). Let $a \otimes b \leq c$. The following inequalities are equivalent:

$$c \multimap f \leq (a \otimes b) \multimap f.$$

By (3.1):

$$c \multimap f \leq b \multimap (a \multimap f).$$

By (RES):

$$(c \multimap f) \otimes a \leq b \multimap f.$$

The second part is proved analogously. Thus, $\mathbf{Q}_j$ is a cyclic nonassociative quantale. □

Let $(\mathbf{Q}, \bigvee, \cdot, \leq)$ be a nonassociative quantale, let $f \in Q$ be a cyclic, shifting element and let $\mathbf{Q}_j$ be the cyclic nonassociative quantale defined in the proof. We define $\multimap$ and   in $\mathbf{Q}$. We show that $\multimap$ and   restricted to j-closed elements are residuals of $\otimes$ in $\mathbf{Q}_j$.

Let $a, b, c \in Q_j$. First, let $a \otimes b \leq c$. Then, $j(a \cdot b) \leq c$. By (QN4) we have $j(a) \cdot j(b) \leq c$, so $a \cdot b \leq c$. By (RES), $b \leq a \multimap c$ and $a \leq c$ b. Conversely, let $b \leq a \multimap c$. Then, $a \otimes b \leq a \otimes (a \multimap c)$. Since c is j-closed, by (QN4'), $a \multimap c$ is j-closed. Hence, $a \otimes b \leq a \otimes (a \multimap c) = j(a \cdot (a \multimap c)) \leq j(c) = c$. The other part is proved analogously.

4 Main Result

In this section, we prove that NCyMALL is a strongly conservative extension of FNL. The idea is straightforward: for an arbitrary set of FNL–sequents (assumptions) and an arbitrary FNL–sequent, we construct a NCyMALL–model in which all assumptions are valid, and the sequent is valid if and only if it is provable in FNL. This model is constructed using nonassociative quantales.

By definition, NCyMALL is a strongly conservative extension of FNL if every translated FNL–sequent is provable in NCyMALL if and only if it is provable in FNL itself. The difficulty lies in the fact that the translation from FNL to NCyMALL alters the structure of sequents, transforming intuitionistic sequents into one-sided sequents. As a result, even translated FNL–sequents may not correspond to proofs that are easily translatable back into the FNL language. Consequently, obtaining a purely syntactic proof becomes challenging. A syntactic proof of the results presented in this section is available in [13]. In this paper, however, we provide a simpler proof using algebraic methods.

The idea is as follows. First, we define a special equivalence relation on sequents, informally described as being insensitive to changes in the order and tree–structure of elements within bunches. We then show that each equivalence class under this relation contains at most one translated FNL–sequent. Therefore, we can work directly with these equivalence classes. Finally, we use this relation to construct suitable models and prove the main theorem.

4.1 Equivalence Relation

We define the relation $\Gamma \sim \Delta$, which holds for the bunches Γ and Δ, if Δ can be derived from Γ by finitely many applications of (r-cyc) and (r-shift).

Since both (r-cyc) and (r-shift) are reversible, this relation is an equivalence relation. The following lemma and its proof follow Buszkowski [5].

Lemma 4.1 *Let $\Gamma[A^{\sim}]$ be an FNL-bunch, containing exactly one occurrence of $A^{\sim}$, where $A^{\sim}$ is a formally negated formula*[3]*. Then, there exists unique Δ such that $\Gamma[A^{\sim}] \sim (A^{\sim}, \Delta)$.*

Proof We provide an algorithm which reduces Γ to some sequent $(A^{\sim}, \Delta)$. The reduction rules are as follows:

$$\begin{array}{ll} \text{(R1)} & (\Phi, \Psi[A^{\sim}]) \multimap (\Psi[A^{\sim}], \Phi) \\ \text{(R2)} & ((\Phi[A^{\sim}], \Psi), \Xi) \multimap (\Phi[A^{\sim}], (\Psi, \Xi)) \\ \text{(R3)} & ((\Phi, \Psi[A^{\sim}]), \Xi) \multimap (\Psi[A^{\sim}], (\Xi, \Phi)) \end{array}$$

(R1) is an application of (r-cyc), (R2) is an application of (r-shift) and (R3) is an application of (r-shift), (r-cyc) and (r-shift), consecutively. The algorithm is deterministic and hence, after finitely many steps, terminates and yields $(A^{\sim}, \Delta)$.

We show, that if Γ reduces to $(A^{\sim}, \Delta)$ and $\Theta \sim \Gamma$, then Θ reduces to $(A^{\sim}, \Delta)$. We use induction on the number of applications of (r-cyc) or (r-shift) needed to obtain Θ from Γ.

Let Θ be obtained from Γ in one application of (r-cyc). If $\Gamma = (\Psi[A^{\sim}], \Phi)$, then $\Theta = (\Phi, \Psi[A^{\sim}])$. We run the algorithm on Θ. The first step is (R1) and we obtain Γ and the algorithm leads to $(A^{\sim}, \Delta)$. If $\Gamma = (\Phi, \Psi[A^{\sim}])$, then $\Theta = (\Psi[A^{\sim}], \Phi)$. We run the algorithm on Γ and from (R1) we obtain Θ and the algorithm leads to $(A^{\sim}, \Delta)$.

Let Θ be obtained from Γ in one application of (r-shift). If $\Gamma = ((\Psi[A^{\sim}], \Phi), \Xi)$, then $\Theta = (\Psi[A^{\sim}], (\Phi, \Xi))$. We run the algorithm on Γ. The first step is (R2) and we obtain Θ, so Θ reduces to $(A^{\sim}, \Delta)$. If $\Gamma = ((\Phi, \Psi[A^{\sim}]), \Xi)$, then $\Theta = (\Phi, (\Psi[A^{\sim}], \Xi))$. We run the algorithm on Θ. The first step is (R1) and we obtain $((\Psi[A^{\sim}], \Xi), \Phi)$. The next step is (R2) and we get $(\Psi[A^{\sim}], (\Xi, \Phi))$. We run the

[3] By a formally negated formula we mean a formula with a superscript $^{\sim}$. This is not a metalanguage negation defined above.

algorithm on Γ. The first step is (R3) and we obtain $(\Psi[A^{\sim}], (\Xi, \Phi))$. So algorithm running either on Γ or Θ leads to $(\Psi[A^{\sim}], (\Xi, \Phi))$. Since the algorithm is deterministic, the result is the same. If $\Gamma = ((\Psi, \Phi), \Xi[A^{\sim}])$, then $\Theta = (\Psi, (\Phi, \Xi[A^{\sim}]))$. We run the algorithm on Γ. We obtain $(\Xi[A^{\sim}], (\Psi, \Phi))$ by (R1). We run the algorithm on Θ and obtain $((\Phi, \Xi[A^{\sim}]), \Psi)$ by (R1) and $(\Xi[A^{\sim}], (\Psi, \Phi))$ by (R3). Again, the algorithm leads to the same on both Γ and Θ.

We assume that Θ is obtained from Γ by $n+1$ applications of (r-cyc) or (r-shift). Let Θ' be obtained from first n applications of the rules on Γ. By induction hypothesis Θ' reduces to $(A^{\sim}, \Delta)$. Also $\Theta' \sim \Theta$ and Θ arises in one step, so Θ reduces to $(A^{\sim}, \Delta)$.

We prove the uniqueness of Δ. Let $\Gamma \sim (A^{\sim}, \Delta)$ and $\Gamma \sim (A^{\sim}, \Delta')$. Then $(A^{\sim}, \Delta) \sim (A^{\sim}, \Delta')$. So $(A^{\sim}, \Delta')$ reduces to $(A^{\sim}, \Delta)$. But the algorithm terminates on the sequents of the form $(A^{\sim}, \Delta)$, so $\Delta = \Delta'$. □

Corollary 4.2 *Let $\Gamma[A^{\sim}]$ reduce to $(A^{\sim}, \Delta)$ and let Θ be a substructure of $\Gamma[A^{\sim}]$, which does not contain $A^{\sim}$. Then, the reduction preserves Θ.*

Proof Let $\Gamma'[A^{\sim}][\Theta] = \Gamma[A^{\sim}]$. We apply Lemma 4.1 to $\Gamma'[A^{\sim}][-]$ and obtain $(A^{\sim}, \Delta'[-])$. Now we substitute Θ in the whole reduction in the place of $-$. Since the algorithm is deterministic, $\Delta = \Delta'[\Theta]$. Hence, Θ is preserved. □

4.2 *The Main Proof*

One notices that $(\otimes \Rightarrow)$ is reversible, so we can represent every sequent $\Gamma \Rightarrow A$ as $G \Rightarrow A$, where G arises from Γ by using $(\otimes \Rightarrow)$ as long as we do not obtain a formula. If $\Gamma = \epsilon$, then $G = 1$.

Theorem 4.3 *Let Φ be a set of FNL-sequents (of the form $C \Rightarrow D$) and let $\Gamma \Rightarrow A$ be an FNL-sequent. Then, $\Phi \vdash_{FNL} \Gamma \Rightarrow A$ iff $\Phi \vdash_{NCyMALL} A^{\sim}, \Gamma$.*

Here, we treat $\multimap$ and as definable symbols (similarly as in the translation). By $A^{\sim}$ we mean the metalanguage negation of A.

Proof We consider the free monoid $(M, \cdot, \epsilon)$ generated by all FNL–formulas and formally negated FNL–formulas. The elements of M are bunches of FNL–formulas and negated FNL–formulas. We define the set $O \subseteq M$ as the smallest set containing all bunches of the form $A^{\sim} \cdot \Gamma$ such that $\Phi \vdash_{\text{FNL}} \Gamma \Rightarrow A$, and closed under (r-cyc) and (r-shift). Each element of O has exactly one negated formula.

Let $\mathbf{Q} = (P(M), \bigcup, \cdot, \subseteq)$, where $X \cdot Y = \{x \cdot y : x \in X, y \in Y\}$. Then $\mathbf{Q}$ is a quantale. The element O is cyclic and shifting in $\mathbf{Q}$, because it is closed under (r-cyc) and (r-shift). We consider cyclic nonassociative quantale $\mathbf{Q}_j$ constructed as in the proof of Proposition 3.8.

We define $[A] = \{\Gamma : \Phi \vdash_{\text{FNL}} \Gamma \Rightarrow A\}$ for any FNL–formula A. One shows that $[A] = \{A^{\sim}\}^{\sim} = \{A^{\sim}\} \to O = \bigcup\{X : \{A^{\sim}\} \cdot X \subseteq O\} = \{\Gamma : A^{\sim} \cdot \Gamma \in O\}$, by the definition of O. Since $[A]$ is of the form $Y^{\sim}$, then $[A]$ is a closed set.

We show the following properties:

(1) if $A \cdot B \in X$, where X is a closed set, then $A \otimes B \in X$ for all FNL–formulas A, B.
(2) if $A \in X$, where X is a closed set, then $[A] \subseteq X$.
(3) if $A \in X$ and $B \in X$, where X is a closed set, then $A \vee B \in X$.

Let $A \cdot B \in X = Y^{\sim}$ and let $\Gamma \in Y$. Then $\Gamma \cdot (A \cdot B) \in O$. Since every element in O has exactly one negated formula, then $\Gamma = \Gamma'[C^{\sim}]$. By Lemma 4.1 and Corollary 4.2, $(\Gamma, (A, B)) \sim (C^{\sim}, \Theta[(A, B)])$. Because O is closed under (r-cyc) and (r-shift), $\Phi \vdash_{\mathrm{FNL}} \Theta[(A, B)] \Rightarrow C$. We use $(\otimes \Rightarrow)$ and get $\Phi \vdash_{\mathrm{FNL}} \Theta[A \otimes B] \Rightarrow C$. Since $(\Gamma, A \otimes B) \sim (C^{\sim}, \Theta[A \otimes B])$, we have $\Gamma \cdot A \otimes B \in O$ and $A \otimes B \in X$. This proves (1).

Let $A \in X = Y^{\sim}$, $\Gamma \in Y$ and $\Delta \in [A]$. Then $\Gamma \cdot A \in O$ and $\Gamma = \Gamma'[C^{\sim}]$. By Lemma 4.1, $(\Gamma, A) \sim (C^{\sim}, \Theta[A])$ and $\Phi \vdash_{\mathrm{FNL}} \Theta[A] \Rightarrow C$. Since $\Phi \vdash_{\mathrm{FNL}} \Delta \Rightarrow A$, we obtain $\Phi \vdash_{\mathrm{FNL}} \Theta[\Delta] \Rightarrow C$ by (cut). Hence, $\Gamma, \Delta \in O$ and $\Delta \in X$. This proves (2).

Let $A \in X$ and $B \in X$, where $X = Y^{\sim}$. Let $\Gamma \in Y$, then $\Gamma \cdot A \in O$ and $\Gamma \cdot B \in O$. We have $\Gamma = \Gamma'[C^{\sim}]$ and by Lemma 4.1 we obtain $(\Gamma, A) \sim (C^{\sim}, \Theta[A])$ and $(\Gamma, B) \sim (C^{\sim}, \Theta[B])$. So, $\Phi \vdash_{\mathrm{FNL}} \Theta[A] \Rightarrow C$ and $\Phi \vdash_{\mathrm{FNL}} \Theta[B] \Rightarrow C$. By $(\vee \Rightarrow)$, we get $\Phi \vdash_{\mathrm{FNL}} \Theta[A \vee B] \Rightarrow C$. Hence, $\Gamma \cdot A \vee B \in O$ and $A \vee B \in X$. This proves (3).

We define a valuation μ of NCyMALL–formulas as follows:

$$\mu(p) = [p], \text{ for every variable } p$$

We recall that $\mu(A) \otimes \mu(B) = j(\mu(A) \cdot \mu(B))$, $\mu(A) \vee \mu(B) = j(\mu(A) \cup \mu(B))$ and $\mu(A) \wedge \mu(B) = \mu(A) \cap \mu(B)$.

We show that $\mu(A) = [A]$ for every FNL–formula by induction on formulas. By the definition of μ, the statement is true for atomic formulas.

1° $A = 1$. One notices:

$$I = j(\{\epsilon\}) = ((\{\epsilon\} \to O) \to O) = O \to O$$

Let $\Gamma \in [1]$, then $\Phi \vdash_{\mathrm{FNL}} \Gamma \Rightarrow 1$. Let $\Delta \in O$, then Δ is equivalent to some $A^{\sim} \cdot \Theta[\epsilon] \in O$ and $\Phi \vdash_{\mathrm{FNL}} \Theta[\epsilon] \Rightarrow A$. By $(1 \Rightarrow)$ we obtain $\Phi \vdash_{\mathrm{FNL}} \Theta[1] \Rightarrow A$ and by (cut) $\Phi \vdash_{\mathrm{FNL}} \Theta[\Gamma] \Rightarrow A$, hence $A^{\sim} \cdot \Theta[\Delta] \in O$ and $\Delta \cdot \Gamma \in O$. So $\Gamma \in O \to O = I$.
Conversely, we have $\{\epsilon\} \subseteq [1]$, hence $j(\{\epsilon\}) = I \subseteq [1]$, since $[1]$ is j-closed.

2° $A = C \otimes D$. By the induction hypothesis, $\mu(C) = [C]$ and $\mu(D) = [D]$. We show $[C \otimes D] \subseteq \mu(C \otimes D)$. We have $\mu(C \otimes D) = \mu(C) \otimes \mu(D) = [C] \otimes [D] = j([C] \cdot [D]) = j(\{(\Gamma, \Delta) : \Gamma \in [C], \Delta \in [D]\})$. Let $\Gamma \in [C]$ and $\Delta \in [D]$. It means that $\Phi \vdash_{\mathrm{FNL}} \Gamma \Rightarrow C$ and $\Phi \vdash_{\mathrm{FNL}} \Delta \Rightarrow D$. By $(\Rightarrow \otimes)$ we have $\Phi \vdash_{\mathrm{FNL}} \Gamma, \Delta \Rightarrow C \otimes D$. This yields $\Gamma \cdot \Delta \in [C \otimes D]$. Since $\Gamma \cdot \Delta \in \{(\Gamma, \Delta) : \Gamma \in [C], \Delta \in [D]\}$ we have $\{(\Gamma, \Delta) : \Gamma \in [C], \Delta \in [D]\} \subseteq [C \otimes D]$. Hence, $\mu(C \otimes D) \subseteq [C \otimes D]$, because $[C \otimes D]$ is j-closed.

We prove $[C \otimes D] \subseteq \mu(C \otimes D)$. We have $C \in [C]$ and $D \in [D]$, so $C \cdot D \in [C] \cdot [D] = \mu(C) \cdot \mu(D) \subseteq \mu(C \otimes D)$. By the properties (1) and (2) we obtain $C \otimes D \in \mu(C \otimes D)$ and $[C \otimes D] \subseteq \mu(C \otimes D)$, since $\mu(C \otimes D)$ is a closed set.

3° $A = C \quad D$. We show $\mu(C \quad D) \subseteq [C \quad D]$. We know that $\mu(C \quad D) \otimes \mu(D) \subseteq \mu(C)$, by (2.1). Then for any $\Gamma \in \mu(C \quad D)$ we know that $\Gamma \cdot D \in \mu(C)$ and, by the induction hypothesis, $\mu(C) = [C]$. By $(\Rightarrow \quad)$ we obtain $\Gamma \in [C \quad D]$.
We show $[C \quad D] \subseteq \mu(C \quad D)$. Let $\Gamma \in [C \quad D]$, i.e., $\Phi \vdash_{\mathrm{FNL}} \Gamma \Rightarrow C \quad D$. We have $\Phi \vdash_{\mathrm{FNL}} C \quad D, D \Rightarrow C$. By (cut) we obtain $\Phi \vdash_{\mathrm{FNL}} \Gamma, D \Rightarrow C$. Hence, $\Gamma \cdot D \in [C]$. Let $\Delta \in [D]$, then again by (cut) we obtain $\Phi \vdash_{\mathrm{FNL}} \Gamma, \Delta \Rightarrow C$, so $\Gamma \cdot \Delta \in [C]$, hence $[C \quad D] \otimes [D] \subseteq [C]$ and finally $[C \quad D] \subseteq [C] \quad [D]$. The case for $C \multimap D$ is similar.

4° $A = C \vee D$. We show $\mu(C \vee D) \subseteq [C \vee D]$. We observe $\mu(C \vee D) = [C] \vee [D] = j([C] \cup [D])$, by definition of μ and the induction hypothesis. Let $\Gamma \in [C] \cup [D]$. Then $\Phi \vdash_{\mathrm{FNL}} \Gamma \Rightarrow C$ or $\Phi \vdash_{\mathrm{FNL}} \Gamma \Rightarrow D$. In both cases we use $(\Rightarrow \vee)$ to obtain $\Phi \vdash_{\mathrm{FNL}} \Gamma \Rightarrow C \vee D$. Hence, $\Gamma \in [C \vee D]$. Since $[C \vee D]$ is j-closed, $\mu(C \vee D) \subseteq [C \vee D]$.
We prove $[C \vee D] \subseteq \mu(C \vee D)$. Since $C \in [C]$ and $D \in [D]$, then $C \in [C] \vee [D]$ and $D \in [C] \vee [D]$. By the induction hypothesis, $[C] \vee [D] = \mu(C) \vee \mu(D) = \mu(C \vee D)$. By the property (3) we obtain $C \vee D \in \mu(C \vee D)$. By the property (2), we get $[C \vee D] \subseteq \mu(C \vee D)$.

5° $A = C \wedge D$. We show $\mu(C \wedge D) \subseteq [C \wedge D]$. Let $\Gamma \in \mu(C \wedge D)$. We observe $\mu(C \wedge D) = [C] \wedge [D] = [C] \cap [D]$, by definition of μ and the induction hypothesis. Then $\Phi \vdash_{\mathrm{FNL}} \Gamma \Rightarrow C$ and $\Phi \vdash_{\mathrm{FNL}} \Gamma \Rightarrow D$, so by $(\Rightarrow \wedge)$ we obtain $\Phi \vdash_{\mathrm{FNL}} \Gamma \Rightarrow C \wedge D$. Hence, $\Gamma \in [C \wedge D]$.
We show $[C \wedge D] \subseteq \mu(C \wedge D)$. Let $\Gamma \in [C \wedge D]$. We have $\Phi \vdash_{\mathrm{FNL}} \Gamma \Rightarrow C \wedge D$ and $\Phi \vdash_{\mathrm{FNL}} C \wedge D \Rightarrow C$ and $\Phi \vdash_{\mathrm{FNL}} C \wedge D \Rightarrow D$. We use (cut) to the respective sequents and obtain that $\Gamma \in [C]$ and $\Gamma \in [D]$. Hence, $[C \wedge D] \subseteq [C] \cap [D]$. By the induction hypothesis and definition of μ, $[C] \cap [D] = \mu(C) \cap \mu(D) = \mu(C \wedge D)$.

Let $G \Rightarrow A \in \Phi$. We show $[G] \subseteq [A]$. Let $\Gamma \in [G]$, then $\Phi \vdash_{\mathrm{FNL}} \Gamma \Rightarrow G$ and by (cut) $\Phi \vdash_{\mathrm{FNL}} \Gamma \Rightarrow A$, so $\Gamma \in [A]$. Thus, $\mu(G) \subseteq \mu(A)$.

We assume $\Phi \nvdash_{\mathrm{FNL}} \Gamma \Rightarrow A$. Then $\Gamma \notin [A]$, by definition. But $\Gamma \in \mu(\Gamma)$ and $[A] = \mu(A)$, so $\mu(\Gamma) \nsubseteq \mu(A)$. Hence, $\Phi \nvdash_{\mathrm{NCyMALL}} A^{\sim}, \Gamma$. □

5 Other Variants

The results of the previous section hold for associative variant of NCyMALL, i.e., CyMALL. We obtain FL and CyMALL from FNL and NCyMALL by replacing bunches with finite sequences of formulas. The rules are similar, but the (r-shift) rule in CyMALL is redundant. The models of FL and CyMALL are the same as for FNL and NCyMALL, but with the assumption that the product is associative.

The analogue of Proposition 3.8 for associative variants is proved by Yetter [14]. One notices that in associative quantales every element is shifting. The analogue of Lemma 4.1 is obvious, since associative (r-cyc) allows cyclic permutations in sequents. In the proof of the main theorem we take $\mathbf{M}_1$ as the free monoid and the rest of the construction of $\mathbf{M}$ remains the same. The rest of the proof may be easily rewritten to fit associative framework.

We may also use fragments of NCyMALL and CyMALL. We consider NCy-MALL and CyMALL without 1. Proposition 3.8 may be proved without unit. In that case f is not a closed element, so it is not an element of the cyclic quantale. The proof of Lemma 4.1 does not require 1 or 0. In the proof of the main theorem, we define $\mathbf{M}_1$ as the free groupoid (or the free semigroup in associative variant). The rest of the construction remains the same. We skip the case when the formula is 1. In other cases we do not use the unit.

We may also consider multiplicative fragments of NCyMALL and CyMALL (with or without 1). In the main proof we skip the cases with additive connectives. Notice that property (3) is used only in the skipped fragment.

References

1. Abrusci VM (1991) Phase semantics and sequent calculus for pure noncommutative classical linear propositional logic. J Symb Log 56(4):1403–1451
2. Abrusci VM (2002) Classical conservative extensions of Lambek calculus. Stud Logica 71(3):277–314
3. Buszkowski W (1982) Some decision problems in the theory of syntactic categories. Math Log Q 28(33–38):539–548
4. Buszkowski, W (2005) Lambek calculus with nonlogical axioms. In: Casadio C, Scott PJ, Seely RAG (eds) Language and grammar. Studies in mathematical linguistics and natural language, pp 77–93
5. Buszkowski W (2016) On classical nonassociative Lambek calculus. In: Amblard M, de Groote P, Pogodalla S, Retoré C (eds) Logical aspects of computational linguistics, Lecture Notes in Computer Science, vol 10054. Springer, pp 68–84. https://doi.org/10.1007/978-3-662-53826-5_5
6. Chvalovský K (2015) Undecidability of consequence relation in full non-associative Lambek calculus. J Symbolic Logic (2):567–586. https://doi.org/10.1017/jsl.2014.39
7. Galatos N, Jipsen P, Kowalski T, Ono H (2007) Residuated lattices: an algebraic glimpse at substructural logics. Elsevier
8. Girard JY (1987) Linear logic. Theoret Comput Sci 50(1):1–101
9. Lambek, J (1958) The mathematics of sentence structure. Am Math Monthly 65(3):154–170. https://doi.org/10.1080/00029890.1958.11989160
10. Lambek J (1961) On the calculus of syntactic types. In: Jakobson R (ed) Structure of language and its mathematical aspects, vol 12. American Mathematical Society, Providence, RI, pp 166–178. https://doi.org/10.1090/psapm/012/9972
11. Mulvey CJ (1986) &. In: Second topology conference, Rendiconti del Circolo Matematico di Palermo. Serie II. Supplemento, vol 12, pp 99–104
12. Niefield SB, Rosenthal KI (1988) Constructing locales from quantales. Math Proc Cambridge Philos Soc 104(2):215–234

13. Płaczek P (2024) Sequent systems for consequence relations of cyclic linear logics. Bull Section Logic 53(2):245–274. https://doi.org/10.18778/0138-0680.2024.06
14. Yetter DN (1990) Quantales and (noncommutative) linear logic. J Symbolic Logic 55(1):41–64. https://doi.org/10.2307/2274953

Representation of Probabilities and Reasoning by Means of Proofs of Linear Logic

Roberto Maieli

Abstract We propose a notion of *probabilistic proof* for the multiplicative and additive fragment of linear logic (MALL) formulated in graphical (*proof-nets*) and sequential style (*derivations*). The definition of probabilistic proof is proved to be correct and complete with respect to the notion of *Bayesian network* that is a graphical model for representing joint probability distributions over a set of discrete (Boolean) random variables. Analogously to what happens with Bayesian networks, probabilistic proofs of linear logic are both an intelligible way of representing probabilities and an efficient method to reason about them.

Keywords Probability distribution · Bayesian network · Sequent calculus · Linear logic · Proof nets

1 Introduction

Linear logic [12] is a (constructive) refinement of classical and intuitionistic logic. Instead of emphasizing *truth*, as in classical logic, linear logic emphasizes the role of formulas as *resources* primarily played within *proofs*. Indeed, one of the most innovative contributions of linear logic is the introduction of two complementary syntaxes for proofs: a sequential one, for expressing proofs as derivation trees in the sequent calculus (*à la* Gentzen, [11]) and a parallel one, for representing proofs as more general graphs, also known as *proof nets*. Given its focus on resources and (mainly) graphical proofs, linear logic has found many applications in Computer Science where programs are theoretically interpreted as "graphs". In this wake, we present here a notion of *probabilistic proof* of linear logic (in *proof nets* and *sequential* style) that allows us to express efficiently both the graphical representation of probability distributions and the reasoning on them, similarly to what happens with Bayesian networks. Bayesian networks [10, 20] are elegant and efficient graphical representations of probability distributions over random (discrete) variables. They are widely

R. Maieli (✉)
Department of Mathematics and Physics, Roma TRE University, Rome, Italy
e-mail: maieli@uniroma3.it

R. Loukanova (ed.), *Computational Linguistics, Information, Reasoning, and AI 2024*, Studies in Computational Intelligence 1280,
https://doi.org/10.1007/978-3-032-21143-9_5

used in Artificial Intelligence and Machine Learning for different purposes such as automatic text classification, spam detection, sentiment analysis, supporting medical diagnosis, etc., Bayesian networks are very useful for inferring (or predicting) the probability of some events given other observed events (evidence). For this reason, it appears natural to try to embed Bayesian networks in logical systems in which it is possible to formalize reasoning on probabilities. Integration of Bayesian networks and first-order logic have been widely investigated (see, e.g., [21]) although it seems quite natural to express Bayesian networks in classical propositional logic. Our proposal is to remain in the propositional fragment of linear logic trying to express efficiently the *graphical representation of probability distributions* via the *proof nets syntax* and, the *reasoning on them*, via the *sequent proofs syntax*. Formally, we set a correspondence between (Boolean) Bayesian networks and a special subclass of proof nets of the multiplicative-additive fragment (MALL, [1, 2, 14–16, 18]) with the Mix rule of the propositional fragment of linear logic: this is called the class of *Bayesian proof nets*. Proof nets are a graphical syntax that allows demonstrations of linear logic to be expressed in a modular and parallel way, abstracting away from the often useless bureaucracy of the sequential calculus. Actually, it is not a novelty to propose variants of proof nets inspired by the methodologies used in Machine Learning or more generally by AI as, e.g., the neural variant of proof nets based on Sinkhorn networks by [13]. The novelty of Bayesian proof nets is given by the fact that they are special proofs endowed with a probability distribution over the set of their "slices". Intuitively, a multiplicative-additive proof net π with conclusions $A_1, ..., A_n$ can be viewed as a "sum" (a superposition) of simpler, multiplicative proof nets, called "slices". Therefore our idea is to associate a probability distribution to the set of slices associated to the proof net π; indeed, since each slice satisfies the invariance of having the same conclusions as π, we can see the probability distribution associated with the slices, as a probability distribution over the conclusions of π itself. Actually, slices play a double role:

- on one hand they allow to detect the "correctness" of the given MALL proof net;
- on the other hand (similarly to what the *chain rule of probabilities* does) they allow to factorize the joint probability distribution associated to the given proof net by means of the (conditional) probability distribution associated to the Bayesian links which the given proof net is built on. The basic idea is that Bayesian links capture the behavior of the discrete random variables of a Bayesian network.

We already know [10, 20] that any Bayesian network $\mathcal{N}$ expresses the joint probability $Pr(A_1, ..., A_n)$ over its n random (discrete) variables $A_1, ..., A_n$. Here we show that any BN $\mathcal{N}$ can be encoded by a probabilistic proof net $\pi^{\mathcal{N}}$ built over n special Bayesian links: exactly one Bayesian link α_i encodes a random variable/node A_i of $\mathcal{N}$; thus, each additive slice $S(\pi^{\mathcal{N}})$ (consisting of an appropriate mutilation of each B-link of $\pi^{\mathcal{N}}$) corresponds to the factorization of the joint probability $Pr(A_1, ..., A_n)$ over the conclusions, $A_1, ..., A_n$, of the proof net $\pi^{\mathcal{N}}$.

Structure of the paper

This paper extends and a further elaborates some ideas illustrated in a previous work presented at DCAI-CLIRAI 2024 [19]. Preliminaries on probabilities for discrete variables are introduced in Sect. 2 while graphical probabilistic models are presented in Sect. 3; in particular, the definition of Bayesian network together with some reasoning techniques on probabilities are discussed in Sect. 3.1 and Sect. 3.2. Then, the graphical counter-part for logical proofs is introduced in Sect. 4 and organized into two subsections: Sect. 4.1 shows the expressive power of the multiplicative and additive fragment of Linear Logic (MALL) and presents the tree-like sequential syntax for logical proofs with the *Mix rule* while the graphical notion of "proof structure" is introduced in Sect. 4.2. Within the MALL fragment we will come across two remarkable sub-fragments: the simple multiplicative fragment of linear logic, known as MLL [9], in Sect. 4.3, and the novelty of the "Bayesian fragment" containing the class of *probabilistic proof nets*, in Sect. 4.4 and Sect. 4.5. Probabilistic proof nets are then showed to be in correspondence with Bayesian networks in Sect. 4.5.3. Conclusions and further works, with possible extensions of our approach to *probabilistic logic programming*, are finally discussed in Sect. 5.

2 Preliminaries on Probability for Discrete Variables

We recall some basic notions on probability for discrete variables. According to the frequentist interpretation, the *probability* of an event from a random experiment is the proportion of the time events of same kind will occur in the long run, when the experiment is repeated, e.g.,

- the probability "my flight to Rome will be on time",
- the probability "this ticket will win the lottery",
- the probability "it will rain tomorrow".

It is always a number in the interval $[0, 1]$ where 0 means "never occurs" and 1 means "always occurs". A *sample space* is a set of possible outcomes for some event. For example, possible instances of sample space for the event "weather tomorrow" are: {rain, not rain} or {sun, rain, snow}, etc.

A *random variable* is a variable representing the outcome of an experiment. For example, if X represents the outcome of the weather tomorrow, we write the probability "tomorrow it will rain" as $Pr(X = \text{rain})$ or, when it's clear which variable we are referring to, we may use the shorthand $P(\text{rain})$. Notation: uppercase letters and capitalized words denote random variables while lowercase letters and un-capitalized words denote values thus, we denote a particular value for a variable as $Pr(X = x)$ or $Pr(Fever = \text{true})$. We also use the shorthand form $Pr(x)$ for $Pr(X = x)$. If X is a random variable, the function $Pr(X = x)$, for each x, is the *probability distribution* of X and it is such that $Pr(x) \geq 0$ for every x and $\sum_x Pr(x) = 1$.

joint probability distribution

x, y	$Pr(X = x, Y = y)$
sun, on-time	**0.20**
rain, on-time	0.20
snow, on-time	0.05
sun, late	**0.10**
rain, late	0.30
snow, late	0.15

marginal distribution

x	$Pr(X = x)$
sun	**0.30**
rain	0.50
snow	0.20

marginal distribution

y	$Pr(Y = y)$
on-time	0.45
late	0.55

conditional distribution

x	$Pr(X = x\|Y = \text{on-time})$
sun	**0.20/0.45=0.444**
rain	0.20/0.45=0.444
snow	0.05/0.45=0.111

Fig. 1 Instances of joint, marginal and conditional distributions

The *joint probability distribution* of two (or more) events is the function given by $Pr(X = x, Y = y)$ (read "X equals x *and* Y equals y"): it returns a number in the interval [0, 1]. Example: the joint probability that "it's sunny and my flight is on time" is given by the first row of the leftmost table of Fig. 1.

The *marginal distribution* of X (i.e., "the distribution of X ignoring other variables") is defined as follows and it is needed to obtain *prior probabilities*:

$$Pr(X) = \sum_{y} Pr(X, y). \tag{1}$$

This definition generalizes to more than two variables, e.g., $Pr(X) = \sum_y \sum_z Pr(X, y, z)$. Example: given the joint probability distribution in the table of Fig. 1, the marginal distribution of "tomorrow will be a sunny day" is given by the first row of the second (starting from the left hand side) table of the Fig. 1:

$$Pr(\text{sun}) = [Pr(\text{sun,on-time}) = 0.20 + Pr(\text{sun,late}) = 0.10] = 0.30.$$

Normalization is the additional step needed to obtain the *conditional probabilities distribution* of X given (the evidence) Y, and defined as "the probability of X given that we know the value of Y"

$$Pr(X = x|Y = y) = \frac{Pr(X = x, Y = y)}{Pr(Y = y)}. \tag{2}$$

Example: "the probability of a sunny day given the evidence that the flight is on time" corresponds to the first row of the rightmost table of Fig. 1 and calculated as follows

$$Pr(X = \text{sun}|Y = \text{on-time}) = \frac{Pr(X{=}\text{sun}, Y{=}\text{on-time}){=}0.20}{Pr(\text{on-time}){=}[Pr(\text{sun,on-time}){+}Pr(\text{rain,on-time}){+}Pr(\text{snow,on-time})]{=}0.45} = 0.444.$$

Two random variables, X and Y, are *independent* if for all x and y, $Pr(x, y) = Pr(x) \times Pr(y)$. Example: X and Y are not independent w.r.t. the joint and marginal distributions given in the tables of Fig. 1 since, e.g., $Pr(\text{sun,on-time}) \neq Pr(\text{sun}) \times Pr(\text{on-time})$.

The *chain rule of probability* allows to factorize the joint probability of two or more variables as follows:

$$Pr(X, Y) = Pr(X|Y) \times Pr(Y) = Pr(Y|X) \times Pr(X)$$

$$Pr(X, Y, Z) = Pr(X|Y, Z) \times Pr(Y|Z) \times Pr(Z) = \frac{Pr(X, Y, Z)}{Pr(Y, Z)} \times \frac{Pr(Y, Z)}{Pr(Z)} \times Pr(Z). \quad (3)$$

In the following we restrict to consider only Boolean random variables: a *Boolean random variable* is a random variable that can take only two possible values, typically represented as *True* and *False* (resp., 1 and 0). It is commonly used in probability theory and statistics to represent outcomes of a binary event, such as "Flipping a coin" (Heads (1) or Tails (0)), "A light switch" (On (1) or Off (0)), "Success or Failure" in an experiment, ect. The main key characteristic of Boolean random variables is given by the probability distribution: any variable is associated with a probability p for one of the values in such a way that if $Pr(X = 1) = p$ then $Pr(X = 0) = (1 - p)$ (or dually, if $Pr(X = 0) = p$ then $Pr(X = 1) = (1 - p)$). Boolean random variables are usually denoted by uppercase letters $A, X...$ while lowercase letters $a, x, ...$ denote *(boolean) values* as follows: a denotes $A =$ True and $\neg a$ denotes $A =$ False. Example: by shorthands, we may write Pr(fever) for Pr(Fever = True) and $Pr(\neg$fever) for Pr(Fever = False). Note that the joint probability distribution $Pr(X, Y)$ of two (or more) Boolean events, X and Y, can be represented as a table (in which each row is the probability of the joint events) or as a (possibly multi-dimensional, in case of many variables) matrix; here we adopt the former formulation (with tables):

(joint probability distribution)

$$Pr(x, y) = Pr(X = \text{True}, Y = \text{True}) = 0.2$$
$$Pr(\neg x, y) = Pr(X = \text{False}, Y = \text{True}) = 0.1$$
$$Pr(x, \neg y) = Pr(X = \text{True}, Y = \text{False}) = 0.3$$
$$Pr(\neg x, \neg y = Pr(X = \text{False}, Y = \text{False}) = 0.6$$

as a table

X	Y	$Pr(X, Y)$
True	True	0.2
False	True	0.1
True	False	0.3
False	False	0.6

as a table

X	Y	$Pr(X, Y)$
x	y	0.2
$\neg x$	y	0.1
x	$\neg y$	0.3
$\neg x$	$\neg y$	0.6

as a matrix

	y	$\neg y$
x	0.2	0.3
$\neg x$	0.1	0.6

3 Graphical Probabilistic Models

The role of graphs in probabilistic and statistical modeling [20] is multiple: they facilitate both representations of joint probability functions and efficient inferences (or reasoning) from observations. Graphical probabilistic models exploit both probability theory and graph theory. Graphs grasp the qualitative part of model: nodes represent events/random variables, edges represent dependencies between them, and conditional (in)dependence can be seen in graph. Probability grasps the quantitative part of model: local information about nodes and its neighbors, the strength of dependency, way of inference, etc.

3.1 Bayesian Networks

Definition 1 (Bayesian network). A *Bayesian Network* (shortly, BN) $\mathcal{N}$ is a triplet $\langle \mathcal{V}, \mathcal{E}, \mathcal{P} \rangle$ where:

1. $\mathcal{V}$ is a (finite) set of variables $X_1, ..., X_n$;
2. $\mathcal{E}$ is a set of arcs, which together with $\mathcal{V}$ constitutes a directed acyclic graph (DAG), $\mathcal{G} = (\mathcal{V}, \mathcal{E})$;
3. $\mathcal{P} = \{Pr(X_i \mid \mathbf{Pa}(X_i)) : X_i \in V\}$ where $\mathbf{Pa}(X_i)$ stands for the set of *parents* of X_i. In words $\mathcal{P}$ is the set the *conditional probabilities* of all variables given their respective parents. Note that, when X_i is a root $\mathbf{Pa}(X_i)$ is empty. In such a case, the expression $Pr(X_i \mid \mathbf{Pa}(X_i))$ simply stands for the *prior probability* of X_i. The conditional probability $Pr(X_i \mid \mathbf{Pa}(X_i))$ of a node X_i is represented as a table, called the *conditional probability table* (shortly, CPT) associated to X_i.

Note that variables (denoted by capital letters X, X_i, Y, Y_i, A, A_i...) of a Bayesian network will be referred as nodes when they are viewed as members of the underlying graph. Also note that the graphical structure of a Bayesian network can be read from the set of the prior and conditional probabilities. So we can use the symbol $\mathcal{N}$ to refer to the *set of prior and conditional probabilities* $\mathcal{P}$ without causing any confusion. We call *Boolean Bayesian network* (shortly, BBN) a BN whose nodes $X_1, ..., X_n$ represent Boolean random variables. In the following we will refer to Bayesian networks as Boolean Bayesian networks unless otherwise indicated.

BNs are models that reduce complexity of the representation of the probabilities distributions and subsequent inferences. Actually, by the *chain rule* (3), the full probability model can be deduced by a gradual decomposition called, *factorization*:

$$\begin{aligned} Pr(X_1, X_2, ..., X_n) &= Pr(X_1) \times Pr(X_2, ..., X_n|X_1) = \\ &= Pr(X_1) \times Pr(X_2|X_1) \times Pr(X_3, ..., X_n|X_1, X_2) = \cdots = \\ &= Pr(X_1) \times Pr(X_2|X_1) \times Pr(X_3|X_1, X_2) \times \cdots \times Pr(X_n|X_1, ..., X_{n-1}). \end{aligned}$$

BNs further simplify this model: all the direct dependencies are taken in to account and all the other (possible) dependencies are ignored. That's the content of the following Bayesian Network Theorem 1.

Theorem 1 (Bayesian network) *If $\mathcal{N}$ is a Bayesian network with $X_1, ..., X_n$ random variables/nodes then*

$$Pr(X_1, ..., X_n) = \prod_{i=1}^{n} Pr(X_i \mid \mathbf{Pa}(X_i)). \tag{4}$$

3.2 Reasoning with Bayesian Networks

BNs are very useful for inferring or predicting the probability of some events from other observed events: given some observation E (also a set of evidence variables) and some target variable Q (also a set of query variables) then $Pr(Q|E)$ is the

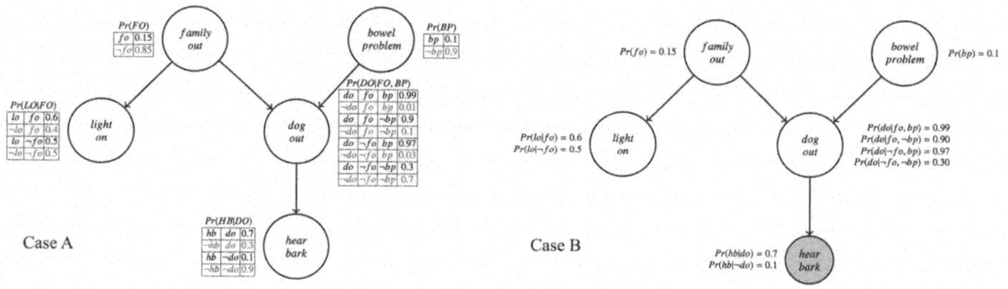

Fig. 2 An instance of Bayesian Network

probability to be found. Assume, e.g., the BN of Fig. 2: both graph and CPTs are from [6] while examples are from [22]; probabilities in grey (Case A) can be omitted, since they can be deduced from the others (Case B).

Now, given the events "no barking heard" ($\neg hb$) and "the door light is on" (lo) what is the probability of the event "family being out", i.e., $Pr(fo|lo, \neg hb)$? Will observation influence the target event? "light on" supports the departure hypothesis while "no barking" suggests the dog is inside ("it is inside" when it is healthy and the family is at home). Let us try to answer the question $Pr(fo|lo, \neg hb)$ via the so called, *inference by enumeration* method. Conditional probabilities are calculated by summing the elements of the joint probability table: but how do we find the joint probabilities if the table is not given?

1. By Theorem 1:

$$Pr(FO, BP, DO, LO, HB) = Pr(FO)Pr(BP)Pr(DO|FO, BP)Pr(LO|FO)Pr(HB|DO);$$

2. by conditional probability definition (2):

$$Pr(fo|lo, \neg hb) = \frac{Pr(fo, lo, \neg hb)}{Pr(lo, \neg hb)}$$

3. then, by probability marginalization (1), we get the numerator:

$$\begin{aligned} Pr(fo, lo, \neg hb) &= \textstyle\sum_{BP,DO} Pr(fo, BP, DO, lo, \neg hb) \\ &= Pr(fo, bp, do, lo, \neg hb) \qquad + Pr(fo, bp, \neg do, lo, \neg hb) \quad + \\ & Pr(fo, \neg bp, do, lo, \neg hb) \qquad + Pr(fo, \neg bp, \neg do, lo, \neg hb) \\ \text{(applying Theorem 1)} &= .033; \end{aligned}$$

4. and similarly, by marginalization, the denominator:

$$Pr(lo, \neg hb) = Pr(fo, lo, \neg hb) + Pr(\neg fo, lo, \neg hb) = .066;$$

5. thus, finally, after substitution:

$$Pr(fo|lo, \neg hb) = \frac{Pr(fo, lo, \neg hb)}{Pr(lo, \neg hb)} = \frac{.033}{.066} = 0.5$$

Therefore, posterior probability $Pr(fo|lo, \neg hb)$ is higher than prior probability $Pr(fo) = 0.15$.

Although in general with BNs we need to express less probabilities than joint probabilities (in the case above, we only needed to express 10 probabilities instead of 2^5 probabilities), enumerating the joint probabilities to answer the query remains a NP-hard problem [7]. Shifting sums left-to-right makes somehow a difference:

$$\begin{aligned} Pr(fo, lo, \neg hb) &= \textstyle\sum_{BP,DO} Pr(fo, BP, DO, lo, \neg hb) \\ &= \textstyle\sum_{BP,DO} Pr(fo)P(BP)Pr(DO|fo, BP)Pr(lo|fo)Pr(\neg hb|DO) \\ &= Pr(fo)\textstyle\sum_{BP} Pr(BB)\sum_{DO} Pr(DO|fo, BP)Pr(lo|fo)Pr(\neg hb|DO) \\ &= Pr(fo)Pr(lo|fo)\textstyle\sum_{BP} Pr(BB)\sum_{DO} Pr(DO|fo, BP)Pr(\neg hb|DO). \end{aligned} \tag{5}$$

Inference by enumeration is an intelligible but inefficient procedure. There are several solutions to optimize the computation of conditional probabilities in BNs as, e.g., the *re-computations minimization* and the *variable elimination procedure* consisting of the following two steps:

1. pre-compute factors to remove some inefficiency (factors allow re-use the earlier computed intermediate results):

$$\sum_P f_1 \times \cdots \times f_k = f_1 \times \cdots \times f_i \times \sum_P f_{i+1} \times \cdots \times f_k$$

   provided that $f_1, ..., f_i$ are factors that do not depend on P;
2. do not consider variables irrelevant to the query: delete all leaves that are neither query nor evidence variable (apply recursively this rule).

Example: try to infer the probability that "the door light is on if the dog is out in the garden", that is:

$$Pr(lo|do) = \frac{Pr(lo, do)}{Pr(do)} = \frac{Pr(lo, do)}{Pr(lo, do) + Pr(\neg lo, do)}.$$

By the enumeration method, we get:

$$\begin{aligned} Pr(lo, do) &= \textstyle\sum_{FO,BP,HB} Pr(FO)Pr(BP)Pr(do|FO, BP)Pr(lo|FO)Pr(HB|do) \\ &= \textstyle\sum_{FO} Pr(FO)Pr(lo|FO)\sum_{BP} Pr(BP)Pr(do|FO, BP)\sum_{HB} Pr(HB|do) \end{aligned}$$

so, by omitting last factor, since $\sum_{HB} Pr(HB|do) = 1$, the factorization is as follows (see also Case B of Fig. 2):

$$Pr(lo, do) = \textstyle\sum_{FO} Pr(FO)Pr(lo|FO)\sum_{BP} Pr(BP)Pr(do|FO, BP). \tag{6}$$

4 Graphical Models of Logical Proofs

In this section we show how BNs can be interpreted as a *probabilistic proofs* of linear logic and how reasoning on BNs corresponds to "slicing" appropriately the corresponding probabilistic proof nets or to deriving the equivalent sequential proofs.

4.1 The MALL with MIX Fragment of Linear Logic

Linear Logic (LL) [12] is a refinement of Classical Logic that has raised a lot of interest in computer science research, especially because of its resource sensitive nature. For our purpose, we restrict here to the pure (without units or neutral elements) *propositional multiplicative and additive fragment* of LL (MALL, briefly). Formulas, denoted A, A_1, ..., B, B_1, ..., are built from *literals*, i.e., propositional variables $a, a^{\perp}, b, b^{\perp}$, ... and by binary connectives $\otimes$ (*tensor*), $⅋$ (*par*), & (*with*) and $\oplus$ (*plus*) according to the following grammar:

$$A, B ::= a, a^{\perp} | A^{\perp} | A \otimes B | A ⅋ B | A \& B | A \oplus B$$

where:

- for conventional reasons, literals $a, a^{\perp}, b, b^{\perp}$, ... are partitioned in two dual groups:

– *positive literals* a, b, ... and
– *negative literals* $a^{\perp}, b^{\perp}$, ...

- *binary connectives* are split in two groups:

– *multiplicative*: conjunction $\otimes$ (*tensor*) and disjunction $⅋$ (*par*),
– *additive*: conjunction & (*with*) and disjunction $\oplus$ (*plus*);

- *negation* $(.)^{\perp}$ is an unary involutive operation ($A^{\perp\perp} = A$) that extends to arbitrary formulas via de Morgan laws:

$$(A \otimes B)^{\perp} = (A^{\perp} ⅋ B^{\perp}) \quad (A ⅋ B)^{\perp} = (A^{\perp} \otimes B^{\perp})$$
$$(A \& B)^{\perp} = (A^{\perp} \oplus B^{\perp}) \quad (A \oplus B)^{\perp} = (A^{\perp} \& B^{\perp}).$$

Observe that in the grammar above we did not introduce the linear implication "$\multimap$" since this connective can be defined via multiplicative disjunction $⅋$ and negation as follows: $A \multimap B =_{def.} A^{\perp} ⅋ B$.

Sequential proofs are derivation trees built according to the inference rules of the $MALL_{mix}$ *Sequent System* defined in Fig. 3 (sequents are one-sided so the turnstile symbol "$\vdash$" may be omitted):

4.2 MALL Proof Structures

One of the main novelties introduced by LL is offered by the *proof nets syntax*. Proof nets are a graphical syntax for proofs, designed in an attempt to abstract away the inessential sequentializations inherent the syntax of sequent systems and defined in

$$\frac{}{a, a^{\perp}}\ ax \qquad \frac{\Gamma, A \qquad \Delta, A^{\perp}}{\Gamma, \Delta}\ cut \qquad \frac{\Gamma \qquad \Delta}{\Gamma, \Delta}\ mix$$

$$\frac{\Gamma, A \qquad B, \Delta}{\Gamma, A \otimes B, \Delta}\ \otimes \qquad \frac{\Gamma, A, B}{\Gamma, A \parr B}\ \parr \qquad \frac{\Gamma, A \qquad \Gamma, B}{\Gamma, A \& B}\ \& \qquad \frac{\Gamma, A_i}{\Gamma, A_1 \oplus_i A_2}\ \oplus_{i=1,2}$$

where Γ, Δ are multi-sets of formulas.

Fig. 3 The $MALL_{mix}$ sequent system

two steps. First, *proof structures* (PS) are defined as directed graphs built by means of *links* (figure below) then, a *correctness criterion* is provided to characterize precisely only those PSs that correspond to effective demonstrations of LL: only correct PSs are called *proof nets* (PNs). PNs quotient classes of sequential derivations that are equivalent up to useless permutations of the inference rules. In the following we introduce a probabilistic variant of proof nets of linear logic.

Definition 2 (proof structure) A *(generalized) MALL proof structure (PS)* π is a directed graph such that each directed edge (an arrow) is labeled by a MALL formula and each node is labeled by a "type" in $\{ax, cut, \otimes^n, \parr^n, \&_a^n, \oplus^n, \bullet\}$ (where $n \geq 1$ is the arity, i.e., the number of premises/entering edges, of the node) and built according to the following typing constraints:

1. entering (resp., exiting) edges of a node L are *premises* (resp., *conclusions*) of L;
2. each edge is conclusion of exactly one node and premise of at most one node;
3. each node $\&_x$ has an eigenvariable $x \in \{a, b, c, ...\}$; multiple $\&$ nodes may share the same eigenvariable x (in that case, they are "copies" of the same $\&_x$-node);
4. "$\bullet$" denotes a dummy node whose premise is also called *conclusion of* π;
5. a node whose conclusion (resp., conclusions) is conclusion (resp. are conclusions) of π is called a *conclusion node* of π.

A *basic link* (or, simply, a *link*) is the pseudo-graph made by a node together with its premise(s) and its conclusion(s) (if any): $\parr$ and $\otimes$ (resp., $\&$ and $\oplus$) links are called *multiplicative links* (resp., *additive links*):

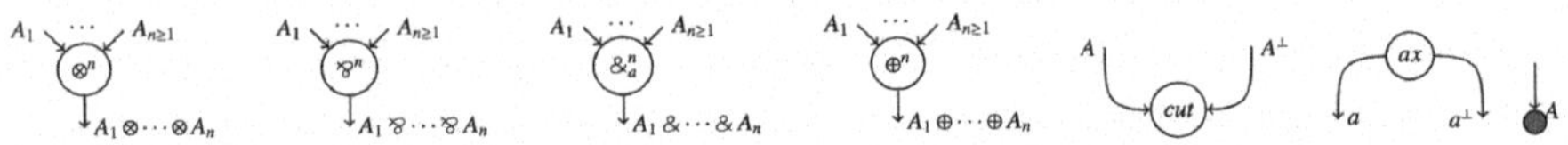

A link with a single (resp., multiple) premise is called unary (resp., binary or n-ary in case $n \geq 2$). Sometimes, unary links (including axioms links) are drawn as simple (possibly, undirected) edges. Additive links with exactly two premises are called *Boolean links*. For our purpose, we do not consider cut links. We say that π is a *multiplicative proof structure* or MLL proof-structure iff π does not contain any additive link ($\&$ or $\oplus_i$). We call, *partial proof structure* (also, *module*), a proof structure with a non-empty set of *pending edges*[1] called the *border*[2] of the structure. A frontier can be partitioned in two parts:

- the *input frontier*, containing only the entering pending edges[3] (if any) and
- the *output frontier*,[4] containing only the exiting pending edges (if any).

Example 1 - Let's "translate" a simple *syllogism* into a derivation of MALL. To simplify, let's assume that A, H, M are literals:

1. "Every Human is Animal": $(H \multimap A) \equiv (H^{\perp} \parr A)$,
2. "Every Human is Mortal": $(H \multimap M) \equiv (H^{\perp} \parr M)$,
3. "Every Human is Animal and Mortal": $(H \multimap (A \,\&\, M)) \equiv (H^{\perp} \parr (A \,\&\, M))$.

We prove that "1 & 2 ⊢ 3" is derivable in MALL sequent system (without Mix rule). First of all, since our calculus is one-sided, we need to move the left-hand side premise "1 & 2" to the right-hand side by applying the linear negation. So, by duality, proving "1 & 2 ⊢ 3" is equivalent to prove "$\vdash (1 \,\&\, 2)^{\perp}, 3$" which, by De Morgan laws, is equivalent to prove "$\vdash 1^{\perp} \oplus 2^{\perp}, 3$" which in turn, by involution of negation, is finally equivalent to prove the sequent $\vdash (H \otimes A^{\perp}) \oplus (H \otimes M^{\perp}), H^{\perp} \parr (A \,\&\, M)$:

$$\dfrac{\dfrac{\dfrac{\dfrac{\dfrac{}{\vdash H, H^{\perp}}\,ax \quad \dfrac{}{\vdash A^{\perp}, A}\,ax}{\vdash H \otimes A^{\perp}, H^{\perp}, A}\,\otimes}{\vdash (H \otimes A^{\perp}) \oplus (H \otimes M^{\perp}), H^{\perp}, A}\,\oplus_1 \quad \dfrac{\dfrac{\dfrac{}{\vdash H, H^{\perp}}\,ax \quad \dfrac{}{\vdash M^{\perp}, M}\,ax}{\vdash H \otimes M^{\perp}, H^{\perp}, M}\,\otimes}{\vdash (H \otimes A^{\perp}) \oplus (H \otimes M^{\perp}), H^{\perp}, M}\,\oplus_2}{\vdash (H \otimes A^{\perp}) \oplus (H \otimes M^{\perp}), H^{\perp}, (A \,\&\, M)}\,\&}{\dfrac{\vdash (H \otimes A^{\perp}) \oplus (H \otimes M^{\perp}), H^{\perp} \parr (A \,\&\, M)}{\vdash ((H^{\perp} \parr A)^{\perp} \oplus (H^{\perp} \parr M))^{\perp}, (H^{\perp} \parr (A \,\&\, M))}\,\equiv \text{(De Morgan + negation)}}\,\parr$$

An alternative compact way of representing proofs of linear logic is through proof structures, which are more general graphs rather than simple trees. Naively, proof structures (without cuts, as in this case) can be constructed in a relatively "simple" manner: first, the syntactic trees of the formulas that are conclusions of the proof are built, and finally, the axiom links are added between pairs of dual literals, taking care of the fact that some literal (leaves) can serve as conclusion of multiple axioms.[5] In such a case, this "sharing" is referred to as "(additive) contraction" (see, e.g., the case of the literal $H^{\perp}$ enclosed in a circle in the proof net below).

[1] An (oriented) edge is said "pending" if exactly one of its end nodes, source or target, is missing.

[2] Also *frontier, interface, layer*.

[3] Edges without source nodes.

[4] Edges without target nodes.

[5] Pairing dual literals "correctly" is not a straightforward task.

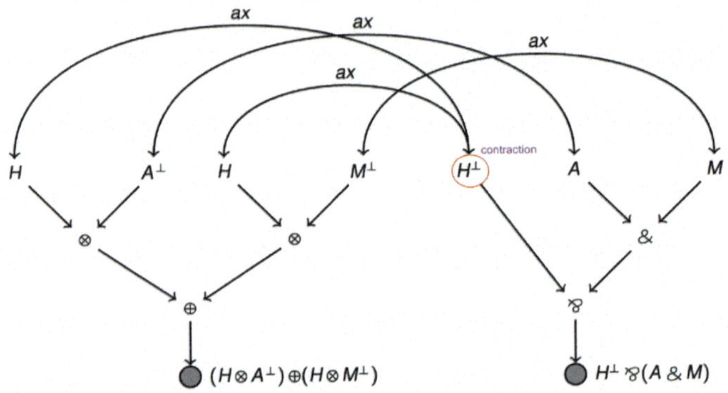

However, not all proof structures that can be constructed on MALL formulas correspond to effective correct proofs of MALL (with possibly Mix): actually, not all MALL formulas are provable. For instance, consider the following multiplicative proof structure which has the formula $a \otimes a^\perp$ as single conclusion: there is indeed no way to prove the sequent $\vdash a \otimes a^\perp$ in the $MALL_{mix}$-sequent system of Fig. 3.

Non-correct proof structure:

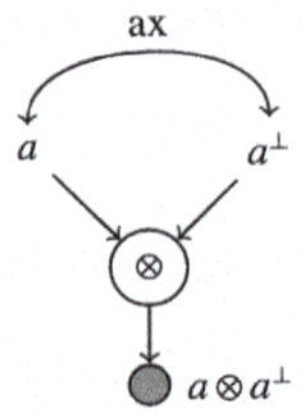

Therefore, we would like structures like this above not to be accepted as admissible structure (i.e., representing proofs). For that reason, in order to characterize only admissible proof structures (those that actually correspond to MALL proofs), we will later introduce the notion of "correctness criterion" and we will call *proof nets* only those proof structures that are *correct* that is, only those structures that are "images", w.r.t. the so called *Sequentialization Theorem* [9], of sequent proofs.

4.3 The Fragment of Multiplicative Linear Logic

The *multiplicative fragment*, MLL, is one of the smallest and most famous fragments of linear logic: it is very expressive and with a plenty of notable results. In addition to the basic rules (axioms and cut) it consists of only two logical rules: the multiplicative conjunction (⊗) and the multiplicative disjunction (⅋) as well as possibly the Mix rule (MLL_{mix}). There is a precise correspondence between the proof nets and the sequential proofs of MLL: this correspondence is established by a famous result known in the literature as the *Sequentialization Theorem* [9]. The most precious ingredient of this theorem is the *Splitting Lemma* which establishes that a correct proof net without conclusions of type ⅋ and with at least one conclusion of type ⊗ is *splitting* that is, there exists a terminal conclusion tensor ⊗-link L such that its removal produces two separated sub-proof nets that are still correct (this is a crucial step to be able to prove the Sequentialization Theorem by induction on the size of a correct proof net). For instance, the proof net below on the left hand side sequentializes in the sequential proof on the right hand side: observe that the leftmost tensor $\otimes_1$ is not splitting, while the rightmost tensor $\otimes_2$ is so; this means that, operationally, $\otimes_1$ cannot be applied as first rule during the bottom-up construction of the sequential proof.

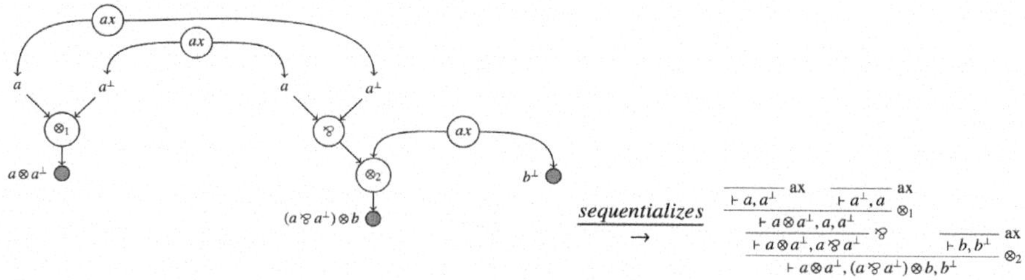

4.4 Bayesian Proof Structures

Bayesian proof structures are particular Boolean MALL proof structures, built by using special "compound links", called *B-links*, obtained by assembling basic multiplicative links together with Boolean additive links and axiom links.[6]

Definition 3 (Bayesian links) A *Bayesian link*, or simply *B-link*, β is a partial proof structure whose border is a (non-empty) multi-set of pairwise dual literals $(a, a^\perp, b_1, b_1^\perp, ..., b_m, ..., b_m^\perp)$. The possibly empty input frontier (resp., output frontier) is at the top (resp., at the bottom) of the B-link. Each B-link has exactly one of the following three forms, as graphically described in Fig. 4:

1. *initial* or *root links* have no $\&$-links and several (at least one) $\oplus$-link possibly grouped by an n-ary $\parr$-link, if $n > 1$. The input interface is empty while the output interface has exactly $n \geq 1$ copies (instances) of the same pair of dual literals $a, a^\perp$ for some a;
2. *internal links* have at least a $\&$-link and at least a $\oplus$-link thus, both the input and the output interfaces are not empty; the input interface is a set of pairwise dual literals, $b_1, b_1^\perp, ..., b_m, b_m^\perp$ (with $b_i \neq b_j$ for all $1 \leq i, j \leq m$) while the output interface contains exactly $n \geq 1$ copies (instances) of the same pair of dual literals, $a, a^\perp$, for some a;
3. *terminal links* consist of a single $\&$-link premise of an unary $\otimes$-link (possibly omitted); its input (resp., output) interface is not empty (resp., empty).

The conclusion of any initial or internal B-link is called the *handle* of the B-link.

---

[6] This notion of compound link is widely used in linear logic and falls within the topic known in the related literature as "generalized connectives" [4, 17].

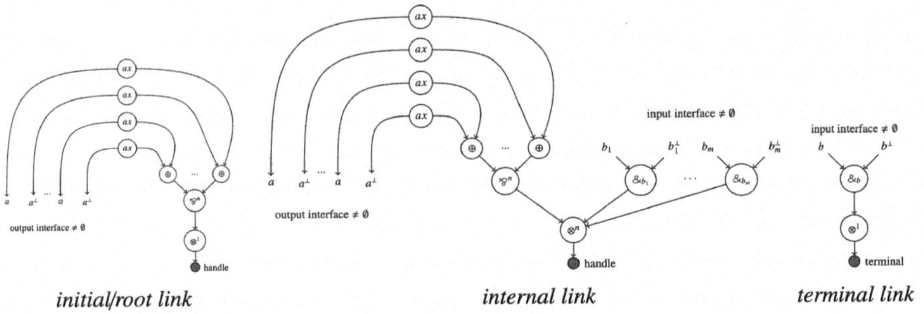

Fig. 4 Bayesian links

Definition 4 (Bayesian proof structure, BPS) A *Bayesian proof structure* (BPS) π is a MALL PS built only using B-links. B-links are juxtaposed through their interfaces: two B-links can be composed by gluing, even partially, their dual interfaces (the input of one B-link with the output of the other B-link) with the proviso that:

- (*duality*) two B-links may share a same literal of their frontiers iff this literal belongs to the input frontier of one link and to the output frontier of the other link;
- (*contraction*) if we glue a B-link whose output frontier consists of $n \geq 1$ occurrences of the same pair of dual literals $a, a^{\perp}$ then, these pairs must be in turn pairwise premises of n copies (instances) of a $\&_a$-link with same eigenvariable a.

Remark 1 Any BPS π satisfies the following facts:

1. since π is a proof structure it does not contain any pending edge so in particular, π has no residual input or output layers;
2. since the input and the output interfaces are sets of pairwise dual literals occurrences, they can only be conclusions of axiom links;
3. all occurrences of $\&$-links and $\oplus$-links in π are conclusions of axioms links;
4. the set of all conclusions of a BPS π coincides with the set of the conclusions of the B-links of π. We call the *effective conclusions*[7] of π only those conclusions of π not belonging to any terminal B-link of π.

In the following we want to precisely characterize only those BPSs that correspond to (or that are "images" of) Bayesian networks (see Theorem 2). Actually, we will show that BNs are DAGs so, reasonably, we want to map any BN into a BPS that has no un-directed cycles. Moreover, not all BPSs correspond to "correct proofs" that is, not all BPSs are images (w.r.t. the so called *sequentialization Theorem*, [12] [9]) of demonstrations of the $MALL_{mix}$ sequent calculus. For these "soundness reasons" we will make use of the notion of "correctness test", usually called *switching* in LL. Naively, we say that a BPS is correct iff it passes a certain number of tests/switchings. Switchings play a double role since they allow:

[7] Only the effective conclusions will be in correspondence with the BN nodes (see Sect. 4.5.1).

1. to detect the correctness of the given BPS and
2. to reason with the conditional probabilities distribution associated to each B-link (thus, allowing the factorization of the joint probabilities over the effective conclusions of BPSs).

Definition 5 (Switchings and Slices) Let π be a BPS. We call *switchable link* any link L of π that is of type $\&$ or $\parr$. Let L be a switchable link of π with $p_1, ..., p_{n\geq 2}$ premises then, a *switch* for L if a (constant) function s that selects exactly one of the n premises of L that is, $s(L) = p_i$ for some $i \in \{1, ..., n\}$ so, there are n different possible switches for L. If L has exactly two premises (it is binary) then s is simply a way to select (or choose) the "Left" or the "Right" premise of L and in that case we simply denote, $s_{\text{left}}(L)$ and $s_{\text{right}}(L)$, the two possible switches of L. In particular, when $L = \&_a$, for some eigenvariable a, then, we may denote $s_{\text{pos}}(L)$ (resp., $s_{\text{neg}}(L)$) the switch that selects the positive premise a (resp., the negative premise $a^{\perp}$) of L.

A *switching* for π, , denoted $S(\pi)$, is a system of switches for all switchable links of π (one switch for each switchable link) with the proviso that all the occurrences of a $\&_a$-link with the same eigenvariable a must be switched in the same way that is, either all "Left" or all "Right" (or equivalently, either all "pos" or all "neg"). Fixed a switching S for π, we call the *switched graph of* π induced by S, denoted $G_S(\pi)$, the undirected graph obtained from π after the following modifications (mutilations, indeed) induced by the system of switches S:

- for every $\parr$-link retain the premise selected by the system S and delete all the other premises (if any);
- for every $\&_a$-link of any eigenvariable a, retain the selected premise together with the axiom link of which this premise is conclusion: select the same positive (resp., negative) premise for all occurrences of the $\&_a$-link indexed by the eigenvariable a and delete the other premise.[8]

A switching can be partial, i.e., it may act only on a subset of switchable links of a proof structure; in particular we call *multiplicative switching* (resp., *additive switching*) a switching that switches only the multiplicative (resp., additive) switchable links.

If π is a MALL PS and S an additive switching for π then we call the *slice* of π induced by S the switched graph $G_S(\pi)$.

Usually, given a proof structure π, the switching systems of π are denoted by $S_1(\pi), S_2(\pi), ..., S_k(\pi), ...,$ etc. Observe that given a MALL PS π and a switching system (resp., an additive switching system) S, the switched graph (resp., the slice) $G_S(\pi)$ is uniquely determined by S so, often, $S(\pi)$ may denote both the switching system and the switched graph (resp., the slice) $G_S(\pi)$ induced by this switching. Observe that any slice is morally a multiplicative proof structure (once the residual unary link are rectified as simple edges). More generally, a proof structure of MALL can be viewed as a super-position of multiplicative (MLL) slices.

---

8 All occurrences of any $\&_a$-link with same eigenvariable "a" are to be considered as multiple "copies" of the same $\&$-node so, they must all be switched in the same way in a fixed swicthing S.

4.5 Bayesian (or Probabilistic) Proof Net (BPN)

Not all Bayesian proof structures correspond to proofs of linear logic and not all proofs correspond to Bayesian networks: only Bayesian proof structure that satisfy a certain criterion and that are endowed with a conditional probability distribution over the conclusions will be in correspondence with Bayesian networks: these structures will be called Bayesian proof nets. Thus, to ensure that a Bayesian proof structure π is also a Bayesian proof net we will proceed in two steps:

1. first we have to check the correctness of π (Sect. 4.5.1),
2. then we have to associate a (conditional) probability table to each non-terminal B-link of π (Sect. 4.5.2).

Compared to the definition of BNs, the first step corresponds to checking the property of being a DAG while the second step corresponds to annotating each variable/node with a conditional probability table (CPT).

4.5.1 Correctness of BPS

In order to define the correctness criterion of proof structure we introduce a graph relation "$\rightsquigarrow$" (also known in the literature on linear logic as "retraction" [8]) that allows to simplify the formulation of the correctness criterion of Bayesian proof structures (Definition 6). We say that a BPS π is *retractible w.r.t. a* iff π contains at least a sub-graph α like that one on the left hand side of the "$\rightsquigarrow$" arrow below, for some eigenvariable a. If π is retractible w.r.t. a then, we say that π *retracts in one step to* π' *w.r.t. a* iff π' is obtained from π by replacing all the occurrences of the sub-graph α on left hand side with the graph β (an axiom link indeed) on the right hand side and we will write $\pi \rightsquigarrow_a \pi'$ where $\pi' = \pi[\beta/\alpha]$.

retraction rule:

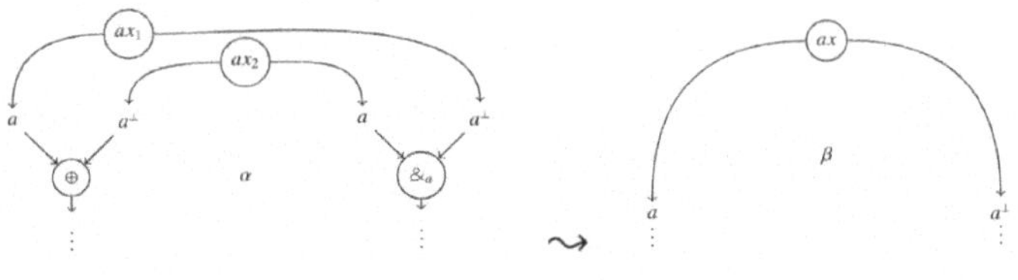

Analogously, we say that π is *rectifiable* iff π contains (at least) one of the two sub-graphs α on the left hand side of the "$\rightsquigarrow$" arrow below. We say that π *rectifies in one step* iff π' is obtained from π by replacing one of the sub-graphs α on left hand side with the graph β (an axiom link indeed) on the right hand side of the "$\rightsquigarrow$" arrow and we will write $\pi \rightsquigarrow \pi'$ where $\pi' = \pi[\beta/\alpha]$.

rectification rules:

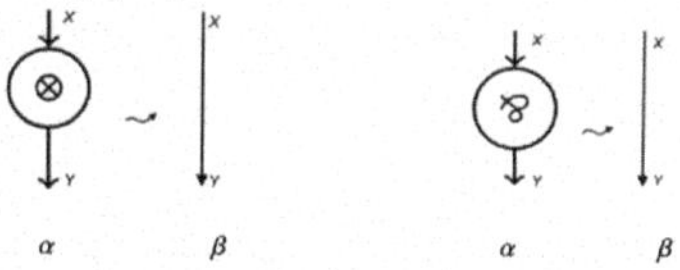

Then we say that π is in *retracted form* (resp., *rectified form*) iff π does not contain any retractible (resp., rectifiable) graph. Moreover, π will be said to be in *normal form* iff π is both in retracted and in rectified form. It is immediate to verify that any BPS in normal form is a multiplicative proof structure. Finally we say that π is *normalizable* iff there exists a finite sequence of retractions or rectification steps terminating with the (unique) normal form. The relation $\rightsquigarrow^*$ on BPS denotes the reflexive and transitive closure of the relation $\rightsquigarrow$.

Definition 6 (Correctness of BPS) Let π be a BPS and let π' be the *normal form* of π. Then, π is a *correct* iff every (multiplicative) switching of π' is an (undirected) acyclic graph.

Note that the technical refinement of retraction and rectification allows the correctness criterion for proof structures to be simplified, reducing it to merely verifying the correctness of a multiplicative proof structure (i.e., only testing the condition of acyclicity of the multiplicative switchings).

Lemma 1 (slices's invariance) *Every correct BPS π satisfies the following facts:*

1. *for every B-link β of π there exists at least a $\&_a$-node in π, for some eigenvariable a, which all the $\oplus$-links that are in the output layer of β depend on (i.e., they will all be switched in the same way);*
2. *if n is the number of $\&$-links of π with different eigenvariable (i.e., we count all links with the same eigenvariable only once) then:*

a. there are exactly n non-terminal B-links in π;
b. there are exactly 2^n slices for π.

proof

1. By (1) of Remark 1 any BPS has no pending edges thus, in particular, all $\oplus$-links occurring in the output layer of a B-link must match with as many occurrences of $\&$-links of as many B-links.
2. Sub-case 1 follows by the construction of BPN.
   Sub-case 1 follows by definition of switching: actually, all occurrences of the $\&_a$-link with same eigenvariable a, are switched in the same way so, it doesn't matter how many there are, there only exit two ways to switch them: either all "Left" or all "Right" (or equivalently, either all "pos" or all "neg").

□

4.5.2 Probabilistic (Bayesian) Proof Nets

To turn a correct Bayesian proof structure π in to a probabilistic proof net we have to add probabilities that is, we have to endow π with a (joint) probability distribution over its n effective conclusions. The reason for excluding terminal B-links is due to the fact that only the non-terminal ones will be "images" (according to Theorem 2) of the variables/nodes of a BN.

Definition 7 (probabilistic proof net) A correct BPS π, with non-terminal links $\beta_1, ..., \beta_n$ and effective conclusions $A_1, ..., A_n$ (handles of, resp., $\beta_1, ..., \beta_n$) is a *probabilistic (or Bayesian) proof net* (shortly, BPN) iff it is equipped with a discrete joint probability distribution over $A_1, ..., A_n$ defined as $\biguplus_{k=1}^{2^n} \langle G_{S_k}(\pi) \rangle$ where each slice $G_{S_k}(\pi)$ takes value $\langle G_{S_k}(\pi) \rangle$ in [0, 1].

Observe that, in the definition above the effective conclusion formulas $A_1, ..., A_n$ are considered as discrete, Boolean indeed, variables; actually, to be precise, for each conclusion A_i there exists one (possibly multiple copies of the) pair of dual literals, a_i and $a_i^\perp$ which denote the two possible values that the variable A_i can take: a_i denotes the value "True" for A_i while $a_i^\perp$ denotes the value "False" for A_i. Now, since in any switching only one of these two literal can occur (the other is deleted) we can consider A_i as a Boolean variable in the sense that A_i can assume only two possible values, a_i and $a_i^\perp$.

Anyway, providing each slice with a joint probability on the proof net's effective conclusions is certainly an intelligible but inefficient way to reason on probabilities. A more efficient way (analogous to that one adopted by BNs) is to assign a conditional probability distributions to the non-terminal B-links of π and then by "slice decomposition" to compute the probability associated to the overall slice of π (a slice of π is a product/composition of the partial slices of the B-links $\beta_1, ..., \beta_n$ building π). This allows us to establish an analogy (a correspondence, indeed) between the factorization (by the chain rule) of a joint probability and the modular decomposition of a proof net slice.[9] Assume π is a correct BPS built over $\beta_1, ..., \beta_s$ B-links with $\beta_1, ..., \beta_{n<s}$ the non terminal ones. Given an additive switching S_k of π, we can calculate the "partial switching" of a single B-link β_i that is, the restriction of $S_k(\pi)$ to β_i which is simply the sub-graph of β_i obtained by appropriately deleting the parts of the graph as indicated by the switching system S_k. Consider, e.g., a generic internal B-link β_i as in Fig. 5: an additive switching $S_k(\pi)$ induces a mutilation, a slice indeed, of β_i (the part of the graph colored in grey is deleted) according to the instructions depicted in Fig. 5:

1. for each $\&_{b_p}$-link of β_i there are two possible switches thus, since $b_p \neq b_q$ for all $p \neq q$ with $1 \leq p, q \leq m$, we have 2^m switches for all m $\&$-links of β_i;
2. for all $\oplus$-links of β_i there are only two possible ways to mutilate them: either all $\oplus$-links are switched Left or all of them are switched Right. This follows from the fact that, by B-links composition, all the occurrences of the $\oplus$-links of

[9] "Modularity" is one of the main advantages of working with a graphical syntax of proofs.

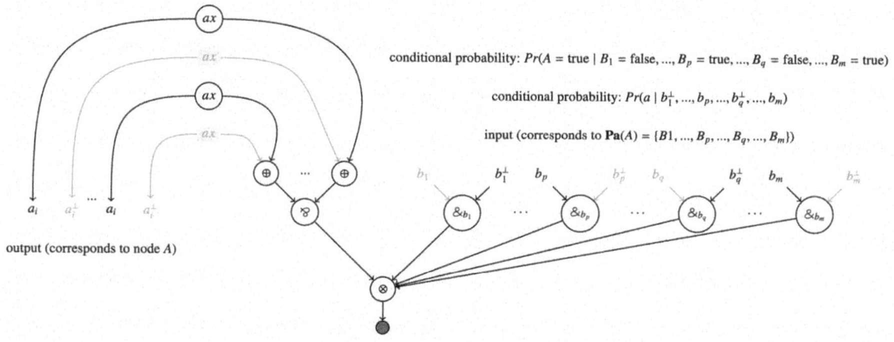

Fig. 5 One of the 2^{m+1} possible "slices" of the B-link β_i

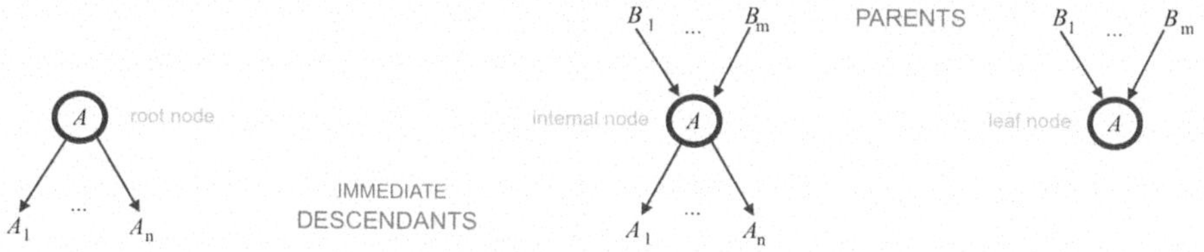

Fig. 6 The three types of random variables/nodes of a Bayesian network

a B-link depend on (i.e., are matched with) as many copies of $\&_a$-link indexed by the same eigenvariable a. Consequently, all the axiom links that are premises of the $\oplus$-links of β_i are all deleted consistently.

From 1 and 2, follows that for the generic internal link β_i of π there are 2^{m+1} possible additive switches (if m is the number of $\&$-links of β_i) while for any initial or terminal B-links there are only two possible switches (regardless of the number of $\oplus$-links). Thus it remains to assign a conditional probability value in the interval [0, 1] to each switched partial graph (the partial slice) $S_K(\beta_i)$. Actually, B-links and the switchings are both inspired by the intuition that any non-terminal B-link β_i encodes a random variable A_i of a BN, with variables $A_1, ..., A_n$, as illustrated in Fig. 6: root nodes are interpreted as root links while internal or leaf nodes are interpreted as internal links as represented in Fig. 4 (this correspondence will be made precise in the next Theorem 2). The input layer of β_i encodes the possible values (True/False) of the parents (variables) nodes of A_i, i.e., $\mathbf{Pa}(A_i) = \{B_1, ..., B_p, ..., B_q, ..., B_m\} \subset \{A_1, ..., A_n\}$ while the output layer encodes the possible values (True/False) of the variable A_i itself, saying whether the immediate descendant nodes of A_i will be active (i.e., value True) or not (i.e., value False) in each row of the conditional probability table. Indeed, the input and the output interfaces of β_i are disjoint sets of pairwise dual literals: m different pairs of literals $b_1, b_1^\perp, ..., b_p, b_p^\perp, ..., b_q, b_q^\perp, ..., b_m, b_m^\perp$ (where m is the number of parents of node A_i) and n instances (copies) of the same pair of literals $a_1, a_1^\perp, ..., a_n, a_n^\perp$ (if n is the number of immediate descendants of A_i). A positive literal b_j (resp. a_i) in the input (resp., output) interface encodes the value

"True" while its dual $b_j^{\perp}$ (resp., $a_i^{\perp}$) encodes the value "False" for the variable B_j that is a parent of A_i (resp., for the variable A_i itself). Remind the basic intuition: the two possible values of any Boolean variable A_i of a BN are "encoded" by (possibly several copies[10] of) a pair of dual literals $a_i, a_i^{\perp}$. Example: consider the slice $S_k(\beta_i)$ of Fig. 5: this figure shows one of the 2^{m+1} possible additive switchings (slices) of an internal B-link β_i (the grey part of the graph is deleted thus, $S_k(\beta_i) \subset \beta_i$). Since in our case β_i is encoding a node A_i whose parents nodes are $B_1, ..., B_m$, the switching $S_k(\beta_i)$ is a way to choose one the 2^{m+1} probability row values of the CPT associated to A_i in the Bayesian network. Actually, each slice selects a literal for each &-link of the input layer and the same literal for all the $\oplus$-links of the output layer, with the proviso that selecting a positive (resp. negative) literal b_i (resp., $b_i^{\perp}$) in the input layer indicates the value True (resp. False) for the corresponding variable $B_i \in \mathbf{Pa}(A_i)$. While selecting the positive (resp. negative) literal in the output layer indicates the value True (resp. False) for the variable itself A_i. So, in particular, the slice $S_k(\beta_i)$ of Fig. 5 is a way to encode the choice of the following probability row value of the CPT of A_i of a corresponding BN:

$$Pr(A_i = \text{true} \mid B_1 = \text{false}, ..., B_p = \text{true}, ..., B_q = \text{false}, ..., B_m = \text{true})$$

which we could also have written as follows (via the linear negation $(.)^{\perp}$):

$$Pr(a \mid b_1^{\perp}, ..., b_p, ..., b_q^{\perp}, ..., b_m).$$

Definition 8 (conditional probability of a slice) Let β_i be a B-link of a correct BPS π and let $S_k(\beta_i)$ be one of the (partial) slices of β_i then, $\langle S_k(\beta_i)\rangle$ denotes the *conditional probability value* associated to $S_k(\beta_i)$ and such that $\biguplus_k \langle S_k(\beta_i)\rangle$ denotes the Conditional Probability Table (CPT) endowing β_i.

From the conditional probabilities associated to the partial slices restricted to every non-terminal B-link β_i we can infer (by factorization) the joint probability of the overall slice of the BPN π which every β_i belongs to. Let π_n be the BPN π restricted to only the non-terminal B-links, $\beta_1, ..., \beta_n$ and let $S_k(\pi_n)$ be a generic partial slice restricted to π_n then, we can compute, by slice decomposition, the joint probability of the overall slice $S_k(\pi_n)$, denoted $\langle S_k(\pi_n)\rangle$, as follows:

$$\begin{aligned}
\langle S_k(\pi_n)\rangle &= \langle S_k(\beta_1 \cdots \beta_n)\rangle && \text{(by BPN construction of}\pi) \\
&= \langle S_k(\beta_1) \cdots S_k(\beta_n)\rangle && \text{(by slice restrictions/decomposition)} \\
&= \langle S_k(\beta_1)\rangle \times \cdots \times \langle S_k(\beta_n)\rangle && \text{(by definition of"}\langle.\rangle\text{", Def.8).}
\end{aligned}$$

Example: given the assumption that every non-terminal B-link β_i of a correct BPS π encodes a random variable A_i of a given BN with variables $A_1, ..., A_n$ then, the joint probability $Pr(A_1 = \text{true}, ..., A_m = \text{true})$ can be encoded by the probability associated to the unique slice $S_{\text{pos}}(\pi_n)$ obtained by setting the switches for all the additive &-links of $\pi = (\beta_1, ..., \beta_s)$ in such a way that only the positive literals are selected and retained while their (negative) duals are erased together with the axioms whose these literals are conclusions of. So, all instances of $a_1, ..., a_n$ are retained while the instances of the negative literals, $a_1^{\perp}, ..., a_n^{\perp}$, are deleted:

---

[10] As many copies as the immediate descendants of A_i.

$$Pr(A_1 = \text{true}, ..., A_n = \text{true}) = Pr(a_1, ..., a_n) \mapsto \langle S_{\text{pos}}(\pi_n)\rangle = \prod_{i=1}^{n} \langle S_{\text{pos}}(\beta_{A_i})\rangle.$$

Therefore, the set of all switchings of π defines a probability distribution over the effective conclusions (random variables) $A_1, ..., A_n$ (or, equivalently, over the non terminal B-links $\beta_{A_1}, ..., \beta_{A_n}$), denoted $\langle \pi_n \rangle$, and defined as follows:

$$\langle \pi_n \rangle = \sum_{k=1}^{n} \langle S_k(\pi_n)\rangle \quad \text{where} \quad \langle S_k(\pi_n)\rangle = \prod_{i=1}^{n} \langle S_k(\beta_{A_i})\rangle.$$

Similarly, reasoning directly on BPNs, we can infer other probabilities as, e.g., the marginal probability. Example: the prior probability $Pr(a_1)$ that, by marginalization, results by the sum $Pr(a_1) = \sum_{A_2} ... \sum_{A_n} Pr(a_1, A_2, ..., A_n)$ over $A_1, ..., A_n$ random variables of a BN, corresponds, in the BPN syntax, to the sum $\sum_{k=1}^{n} \langle S_k^{a_1}(\pi)\rangle$ where $S_k^{a_1}(\pi)$ denotes a generic slice for π in which we always retain all the instances of the positive literal a_1 (i.e., we always delete all the instances of the negative literal $a_1^{\perp}$). Remind that the probability distributions are assigned only to the switching sets of non-terminal links; these are in fact the only links that are in correspondence with the nodes of BNs. Therefore any correct BPS π in which each non terminal B-link β_i is equipped with a conditional probability table in which each row indicates the probability value $\langle S_k(\beta_i)\rangle$ assigned to a slice of $S_k(\beta_i)$ of β_i is called a Bayesian (or probabilistic) Proof Net (Definition 7).

The correspondence between BPNs and BNs is established in the next section.

4.5.3 Encoding BNs in to BPNs

In this section we prove that the notion on Bayesian Proof Net (Definition 7) is correct (Theorem 2) and complete (Theorem 3) w.r.t. the notion of Bayesian Network (Definition 1).

Theorem 2 (Correctness of BPNs) *Given a Boolean Bayesian Network $\mathcal{N}$ with random Boolean variables $A_1, ..., A_n$, there exists a BPN $\pi^{\mathcal{N}}$ with non-terminal B-links $\beta_1, ..., \beta_n$, satisfying the following conditions (we say that $\pi^{\mathcal{N}}$ is an* an encoding *of $\mathcal{N}$):*

1. *for each variable A_i there exactly exists in $\pi^{\mathcal{N}}$ one corresponding non-terminal B-link β_i whose input layer* encodes *the parents nodes $\mathbf{Pa}(A_i)$ of A_i as follows: if B_j is in $\mathbf{Pa}(A_i)$ then there exactly exists in the input layer of β_i a pair of dual literals, b_j and $b_j^{\perp}$, that are premises of a $\&_{b_j}$-link with eigenvariable b_j;*
2. *the table row values of the CPT endowing each A_i in $\mathcal{N}$, are mapped in to $\biguplus_{k=1}^{n} \langle S_k(\beta_i)\rangle$ in the following way: the i-th table row value v_i is assigned to the slice $S_k(\beta_i)$ (i.e., $\langle S_k(\beta_i)\rangle := v_i$) iff:*

a. *the positive literal b_j (resp., the negative literal $b_j^{\perp}$) of the input border of β_i is selected/retained by $S_k(\beta_i)$ iff the variable/node $B_i \in \mathbf{Pa}(A_i)$ has value True (resp. False) in the i-th table row of the CPT with probability value v_i;*

b. *the positive literal a_i (resp., the negative literal $a_i^{\perp}$) of the output border is selected/retained by $S_k(\beta_i)$ iff the variable/node A_i has value True (resp. False) in the i-th table row of the CPT with probability value v_i.*

Proof By induction on the size of $\mathcal{N}$ (i.e., the number of nodes/variables of $\mathcal{N}$).

Base of induction. If $\mathcal{N}$ consists of only one variable A then $\mathcal{N}$ is encoded by the following BPN $\pi^{\mathcal{N}}$ in a such way that the probability distribution associated to A, $Pr(A = \text{true}) + Pr(A = \text{false}) = 1$, corresponds to the probability distribution assigned to the positive (or left) and negative (or right) slice of $\pi^{\mathcal{N}}$ that is, $\langle \pi^{\mathcal{N}} \rangle = \langle S_{pos}(\pi^{\mathcal{N}}) \rangle + \langle S_{neg}(\pi^{\mathcal{N}}) \rangle = 1$.

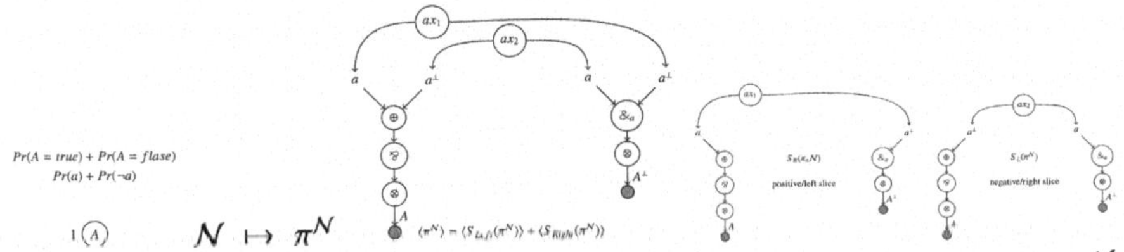

If we ignore (by rectification) the unary multiplicative links (⅋ and ⊗) of $\pi^{\mathcal{N}}$, which have no correspondence in the sequent calculus rules,[11] we can write down the sequent proof corresponding to the BPN $\pi^{\mathcal{N}}$ above and so to the given BN $\mathcal{N}$:

$$\dfrac{\dfrac{\dfrac{}{\vdash a^{\perp}, a}\,ax_2}{\vdash a \oplus a^{\perp}, a}\,\oplus_2 \quad \dfrac{\dfrac{}{\vdash a, a^{\perp}}\,ax_1}{\vdash a \oplus a^{\perp}, a^{\perp}}\,\oplus_1}{\vdash a \oplus a^{\perp}, a \,\&\, a^{\perp}}\,\&$$

Note that, the B-link with handle formula $A = a \oplus a^{\perp}$ is the one encoding (it is image of) the variable node A of $\mathcal{N}$ while, by duality, $A^{\perp} = a \,\&\, a^{\perp}$ is the handle of the terminal B-link.[12]

Induction step. If $\mathcal{N}$ is not reduced to a single node/variable then, since $\mathcal{N}$ is a DAG (so, it does not contain directed cycles), there exists at least a node A that is root and parent of some (at least one) nodes in $\mathcal{N}$, as, e.g., nodes B and C, below:

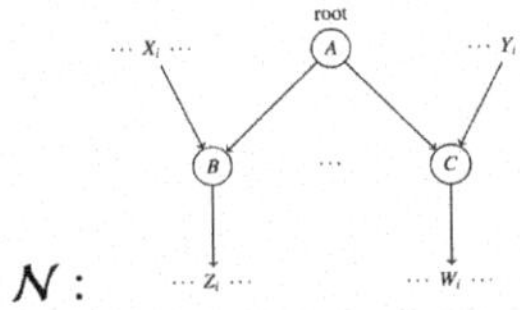

There are two cases depending on whether or not the removal of A splits $\mathcal{N}$ into two (or more, depending on the number of immediate descendants of A) BNs, that is:

- Case 1: the removal of A splits $\mathcal{N}$ in two (or more) BNs, $\mathcal{N}_1$ and $\mathcal{N}_2$;
- Case 2: the removal of A induces a single sub-BN $\mathcal{N}'$.

---

[11] Unary links are used in the graphical syntax to only make consistent the definition of B-link in the general n-ary case ($n \geq 1$): they can always be trivially rectified as simple edges.

[12] We chose the handle formula $a \oplus a^{\perp}$ to be the image of the single node Bayesian network but we could also have made the dual choice, taking $a \,\&\, a^{\perp}$ as the image: that's duality of linear logic.

(Case 1) $\mathcal{N}_1$: 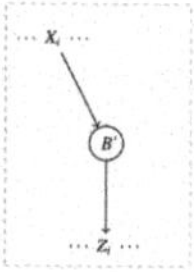 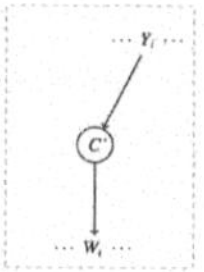 : $\mathcal{N}_2$ (Case 2) $\mathcal{N}'$: 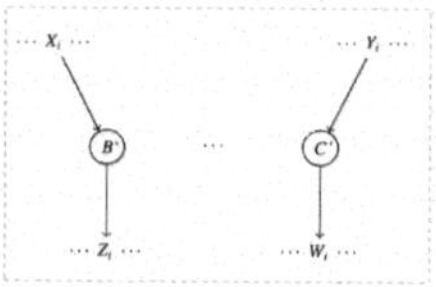

In both cases we can reason by applying the hypothesis of induction. Let's assume we are in Case 1 (the other cases are similar and so omitted), thus by hypothesis of induction on the size, we get two correct BPNs, $\pi^{\mathcal{N}_1}$ and $\pi^{\mathcal{N}_2}$, as follows

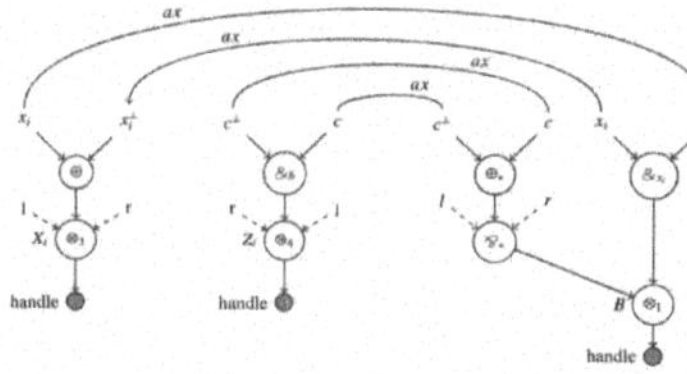

$\pi^{\mathcal{N}_1}$:

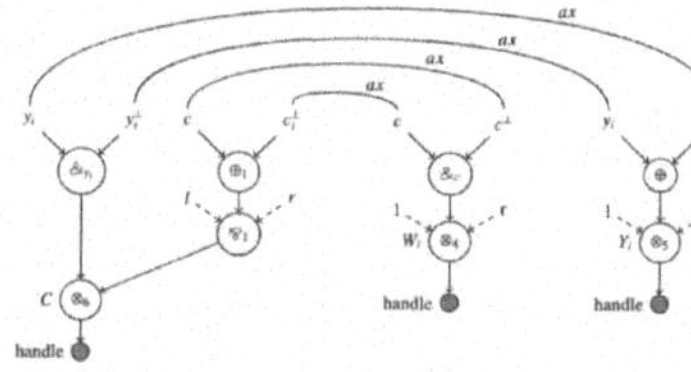

$\pi^{\mathcal{N}_2}$:

that we can merge together, by adding the B-link β_A (image of the node A) below

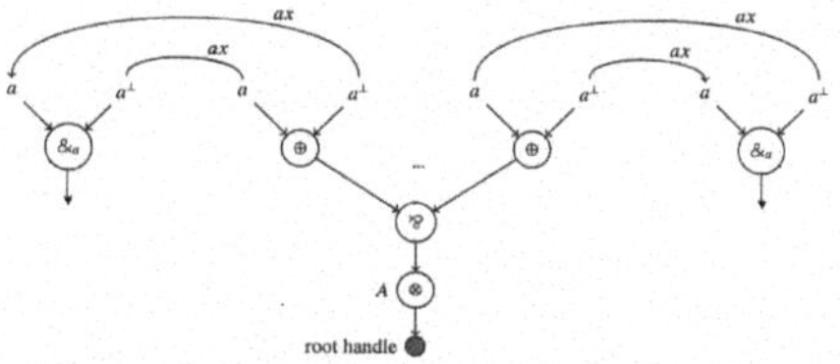

in such a way to get the following connected graph $\pi^{\mathcal{N}}$:

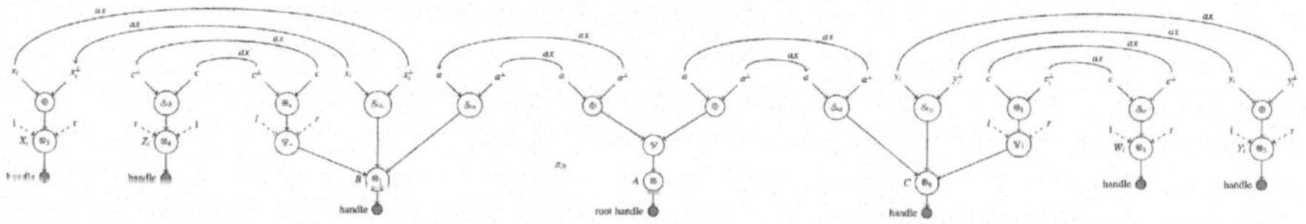

By construction, $\pi^{\mathcal{N}}$ is a Bayesian proof structure. It remains to show that $\pi^{\mathcal{N}}$ is also a correct Bayesian proof net. So, we first reduce $\pi^{\mathcal{N}}$ to the (retracted and rectified) normal form $\pi_r^{\mathcal{N}}$ which is a multiplicative proof structure indeed:

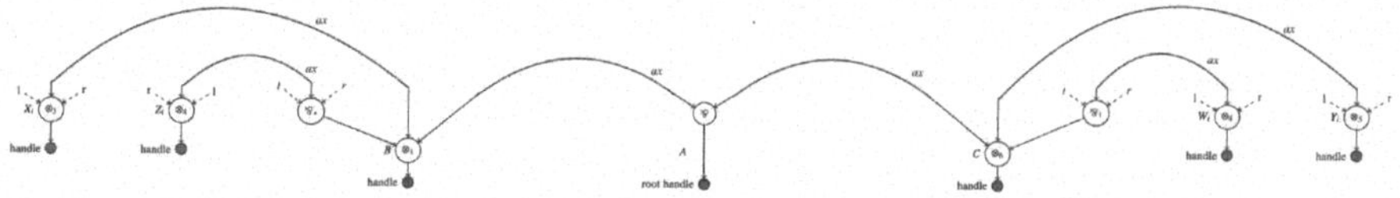

Clearly, every (multiplicative) switching S_k for $\pi^{\mathcal{N}}$ induces an acyclic graph therefore, by Definition 6, $\pi^{\mathcal{N}}$ is a correct BPS otherwise, assume by absurdum that there exists a switching S_k for $\pi_r^{\mathcal{N}}$ containing a cycle. Obviously this cycle cannot pass trough the middle node $\parr$ (the one above the root handle A) since by definition of switching this node is switched so, this cycle must be contained in $\pi^{\mathcal{N}_1}$ of $\pi^{\mathcal{N}_2}$, contradicting their correctness (assumed by hypothesis of induction).

It remains to show that $\pi^{\mathcal{N}}$ is a BPN equipped with an appropriate joint probability distribution over the conclusions $A_1, .., A_n$. We already know, by hypothesys of

induction that every non-terminal B-link β_i of $\pi^{\mathcal{N}_1}$ (resp., $\pi^{\mathcal{N}_2}$) is equipped with a conditional probability distribution $\biguplus_{k=1}^{n}\langle S_k(\beta_i)\rangle$ corresponding to the CPT endowing the variable A_i of $\mathcal{N}_1$ (resp., $\mathcal{N}_2$), thus it remains to assign each of the two row values of the CPT of A to each slice of the B-link β_A (encoding A):
- the row value $Pr(A = \text{True}) = p$ is assigned to the left/positive slice of β_A and
- the row value $Pr(A = \text{False}) = 1 - p$ is assigned to the right/negative slice of β_A:

$Pr(A = \text{true}) = Pr(a) = p \mapsto \langle S_{pos}(\beta_A)\rangle$
$Pr(A = false) = Pr(\neg a) = (1 - p) \mapsto \langle S_{neg}(\beta_A)\rangle$

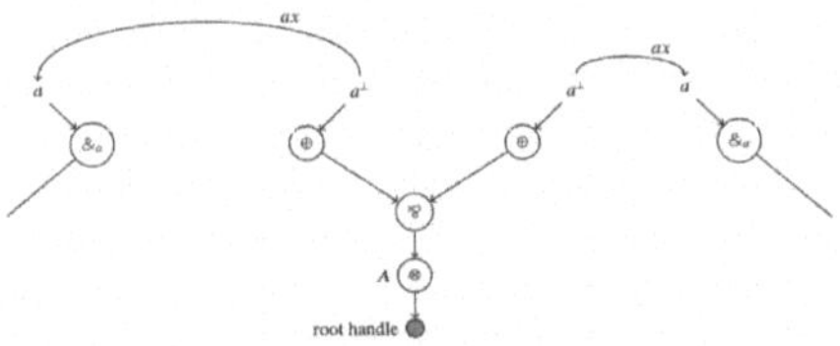

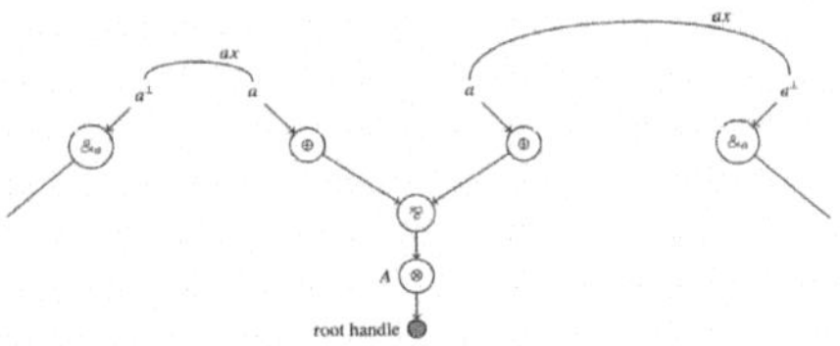

$Pr(a) = p \mapsto \langle S_{pos}(\beta_A)\rangle$ $\qquad$ $Pr(\neg a) = (1 - p) \mapsto \langle S_{neg}(\beta_A)\rangle$ □

Lemma 2 (root link) *Every correct BPS π contains at least a root B-link.*

Proof We reason by absurdum, assuming that there exists at least a correct BPS π that has no root B-link and we show a contradiction. So, let us assume π is a minimal (w.r.t. the size of the graph or the number of B-links) Bayesian PN without root links. Let's start by retracting and rectifying the BPN π until it becomes the proof structure in normal form π^* (in which all the unary $\otimes$ and $\parr$ links are now simple edges). By construction, the normal form π^* is a multiplicative proof structure which is correct that is, every (multiplicative) switching for π^* is an acyclic graph. Observe that since by assumption the Bayesian proof net π has no root link then, after the retraction and rectification steps, all tensors in π^* have at least two premises as in the next l.h.s. picture; moreover, since we assumed that there is no root link, there must exist at least one such (at least binary) tensor, otherwise π^* would consist of a single $\parr$-link and so the initial π would contain a root link. For similar reasons, observe that since there are no unary tensor links in π^* then, there cannot exist in π^* n-ary $\parr$-links that are conclusion links otherwise, there would already exist a root link in π (see the r.h.s. picture below).

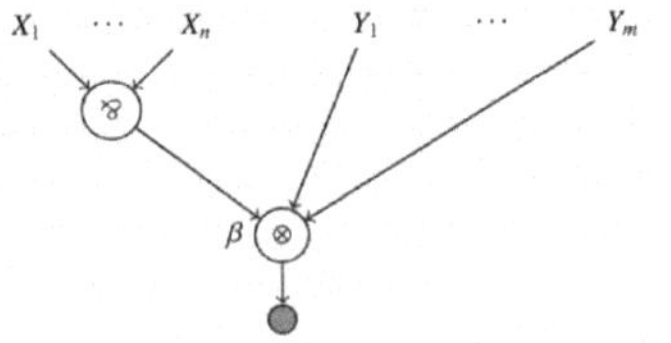

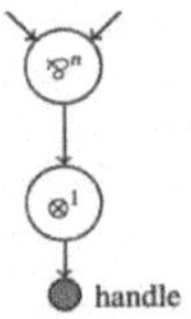

So, by the Splitting Lemma[13] there exists at least one splitting tensor/handle such that its removal splits π^* into two smaller MLL proof structures, π_1^* and π_2^*, that are correct w.r.t. the MLL_{Mix} fragment:

[13] The Splitting Lemma is crucial for the Sequentialization Theorem allowing to apply the induction.

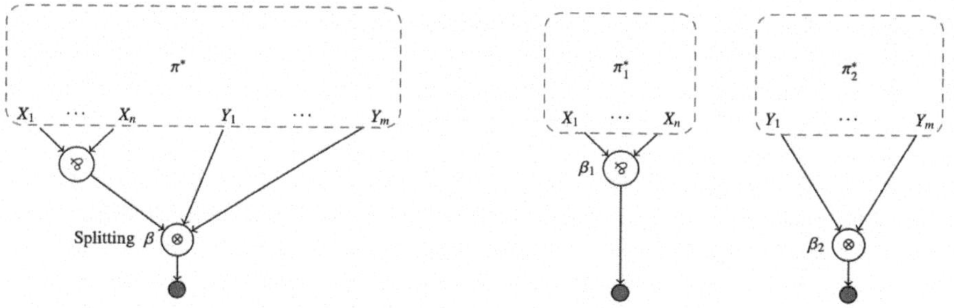

Now, if we restore in π_1^* and π_2^* all the retraction and rectification steps they concerned, we get two correct Bayesian proof nets, π_1 and π_2, which have strictly smaller sizes than π so, for π_1 and π_2 there must exist, resp., at least a root B-link: let us assume that ν_1 and ν_2 are respectively, the root links of π_1 and π_2. In particular, observe that ν_1 may coincide with the restored β_1 in π_1 since β_1 is certainly a root node (after restoring the retraction and rectification steps) in π_1 while β_2 cannot be a root node otherwise β itself would already be a root note in π contradicting the assumption that π has no root nodes. Now, let us restore the splitting tensor removed just before (the application of the Splitting Lemma) in a such a way to get back β and so π. Clearly, since ν_2 is different from β_2 and so from both β_1 and β then, ν_2 would already be a root B-link for π, a contradiction. □

Theorem 3 (BPNs completeness) *For every BPN π there exists a BN $\mathcal{N}$ such that π is an encoding*[14] *of $\mathcal{N}$.*

Proof Analogously to the proof of the Correctness Theorem 2, we reason by induction on the number of non-terminal B-links of the given BPN π.

Base of induction. If π contains a single non-terminal B-link, then both π and $\mathcal{N}$ are like $\pi^{\mathcal{N}}$ and $\mathcal{N}$ in the proof of the *base of the induction* case of Theorem 2.

Induction step. If π contains at least two non-terminal B-links then, by Lemma 2, there exists at least a root B-link, so we proceed by removing a root B-link β from π and then applying the induction hypothesis to $\pi' = \pi \setminus \{\beta\}$ in case π' is a connected graph, otherwise we apply the hypothesis of induction to π_1, π_2 in case that the removal splits the residual graph in two parts. We then conclude reasoning in a similar way to what was done in the *induction step* case of Theorem 2.

It is finally trivial to reconstruct the CPT of any node X_i of the target BN, starting from the conditional probabilities associated to the slices of the ith B-link β_{X_i} (Definition 8) that is in correspondence with the variable X_i (following the *encoding* defined in Theorem 2). □

4.5.4 Reasoning on Probabilities via Proof Nets and Sequential Proofs

In this section, we give some examples showing how to exploit the two syntaxes of proofs of linear logic to represent Bayesian networks (via proof nets) and to reason about them (via sequent proofs).

---

[14] The encoding refers to conditions 1 and 2 of Theorem 2.

Example 2 Consider the encoding of the BN $\mathcal{N}$ below

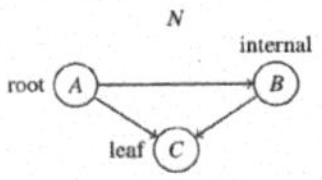

that, via Theorem 2, returns the following BPN $\pi^{\mathcal{N}}$ in which the three non-terminal B-links are enclosed within (colored) dotted lines: β_A (red), β_B (blue) and β_C (green):

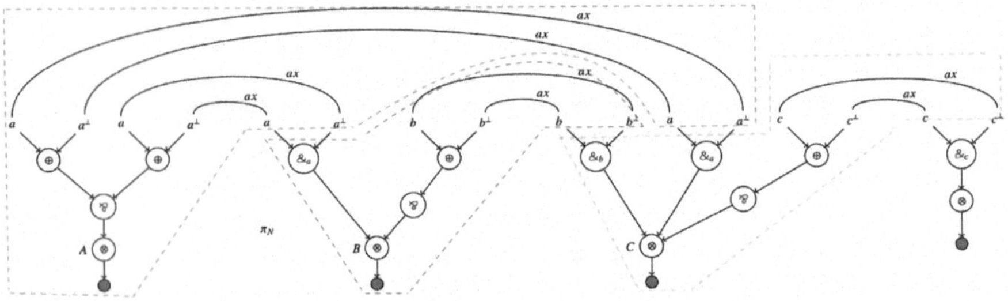

Clearly, this is a correct Bayesian proof structure: actually, its normal form is such that every (multiplicative) switching does not contain cycles (otherwise there is a contradiction with $\mathcal{N}$ being a DAG). Now, observe that given the BPN associated to a BN it is very simple to reason on probabilities working directly on the BPN side. Try, e.g., to compute the joint probability $Pr(a^{\perp}, b, c)$. This probability corresponds to the probability $\langle S_k(\pi_3^{\mathcal{N}})\rangle$ associated to the slice $S_k(\pi_3^{\mathcal{N}})$ in which we retained:

- the right (negative) premise $a^{\perp}$ of (both) the $\&_a$-links (thus all the occurrences of the positive literal a together with the axiom links that have these literals as conclusions are erased) and
- the left premises, b and c, of resp., $\&_b$ and $\&_c$ links (thus both $b^{\perp}$ and $c^{\perp}$, together with the axioms links having these literals as conclusions, are erased):

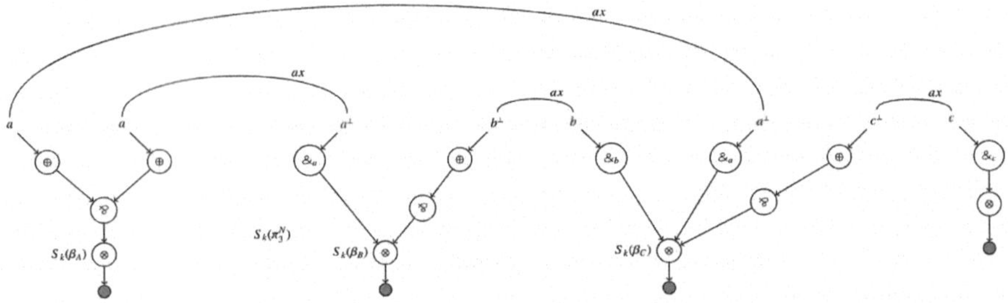

$$\begin{aligned} Pr(a^{\perp}, b, c) \mapsto \langle S_k(\pi_3^{N})\rangle &= \langle S_k(\beta_C \cdot \beta_B \cdot \beta_A)\rangle && \text{(by BPN construction)}\\ &= \langle S_k(\beta_C) \cdot S_k(\beta_B) \cdot S_k(\beta_A)\rangle && \text{(by slice restrictions)}\\ &= \langle S_k(\beta_C)\rangle \cdot \langle S_k(\beta_B)\rangle \cdot \langle S_k(\beta_A)\rangle && \text{(by"}\langle.\rangle\text{", Def.8)}\\ &= Pr(c|b, a^{\perp})Pr(b|a^{\perp})Pr(a^{\perp}) && \text{(by interpret. with cond. prob.).} \end{aligned}$$

While joint probabilities can be computed via *slice factorization* (simulating the chain rule of conditional probabilities), marginal probabilities can be inferred by *summing slices*. Example: summing the slice $S_k(\pi^{\mathcal{N}})$ above with $S_{k'}(\pi^{\mathcal{N}})$ below

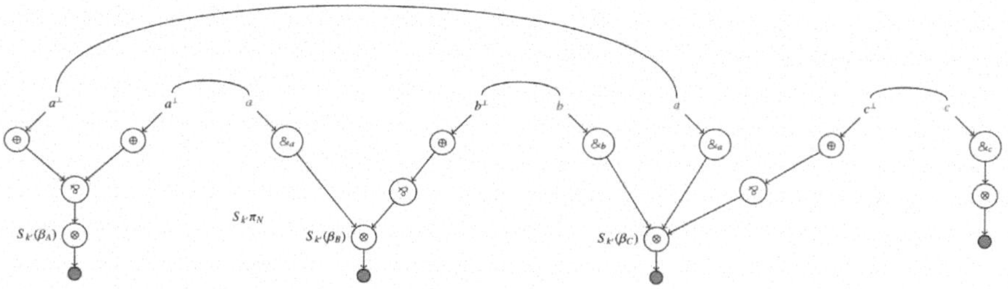

returns the following marginal probability:

$$\begin{aligned} Pr(b,c) = \sum_A Pr(A,b,c) &= Pr(a,b,c) + Pr(a^\perp,b,c) \\ &\mapsto \langle S_k(\pi^{\mathcal{N}})\rangle + \langle S_{k'}(\pi^{\mathcal{N}})\rangle. \end{aligned}$$

Anyway, since there exists a *correspondence*[15] between proof nets and sequential proofs of linear logic, the Bayesian proof net above might as well have been written in sequential style as a proof of the sequent $\vdash \beta_A, \beta_B, \beta_C, G$ (where $G = c \,\&\, c^\perp$ is the conclusion of the rightmost hand side terminal link) of the $MALL_{Mix}$ Sequent System in Fig. 3. The advantage of the sequential syntax is that we can adapt the proof to be written in a *pseudo-logic programming style*. In this way, a Bayesian network (with variables A_1, ..., A_n) will correspond to a derivation of $\vdash \mathcal{P}, \mathcal{G}$, where $\mathcal{P}$ is a *Bayesian program* (i.e., a collection of *Bayesian methods* corresponding to the nodes of a BN) and $\mathcal{G}$ is a non-empty set of *goals* corresponding to (the duals of) the terminal/leaf nodes of the BN. So, in our case, the Bayesian program (*specification* of the BN $\mathcal{N}$ above) is given by the set of *(Bayesian) methods* $\mathcal{P} = \{\beta_A, \beta_B, \beta_C\}$ and the goal set $\mathcal{G} = \{G\}$ below:

$$\begin{aligned} \text{fact: } \beta_A &= (a \oplus a^\perp) \parr (a \oplus a^\perp) \\ \text{method: } \beta_B &= (b \oplus b^\perp) \otimes (a \,\&\, a^\perp) \\ \text{method: } \beta_C &= (c \oplus c^\perp) \otimes (b \,\&\, b^\perp) \otimes (a \,\&\, a^\perp) \\ \text{goal: } G &= (c \,\&\, c^\perp) \end{aligned}$$

where a *Bayesian method* is any MALL formula with the following general pattern

$$((a \oplus a^\perp) \parr \cdots \parr (a \oplus a^\perp)) \otimes ((b_1 \,\&\, b_1^\perp) \otimes \cdots \otimes (b_n \,\&\, b_n^\perp))$$

in which, resp., $((a \oplus a^\perp) \parr \cdots \parr (a \oplus a^\perp))$ is the *head* and $((b_1 \,\&\, b_1^\perp) \otimes \cdots \otimes (b_n \,\&\, b_n^\perp))$, with $n \geq 0$, is the *body* of the method. In case of body vanishing (i.e., $n = 0$) then the method is called a *fact*. Observe that the head is a generalized $\parr$ connective whose arguments are (possibly) several copies of the same pair of dual literals, $a, a^\perp$. Analogously, in case of head vanishing, methods are called *goals*. In general, $\mathcal{G}$ denotes a non-empty set of goals. Finally,

[15] This correspondence is established by the *Sequentialization Theorem* ([12] and [9]).

the proof of the sequent $\vdash \beta_A, \beta_B, \beta_C, G$, in a pseudo-logic programming style, results as follows[16]:

$$\dfrac{\dfrac{\dfrac{\dfrac{\dfrac{}{\vdash c^{\perp},c}\,ax}{\vdash c\oplus c^{\perp},c}\,\oplus_2 \quad \dfrac{\dfrac{}{\vdash c,c^{\perp}}\,ax}{\vdash c\oplus c,c^{\perp}}\,\oplus_1}{\vdash c\oplus c^{\perp},c\,\&\,c^{\perp}}\,\& \quad \dfrac{\dfrac{\dfrac{}{\vdash a^{\perp},a}\,ax}{\vdash a\oplus a^{\perp},a}\,\oplus_2 \quad \dfrac{\dfrac{}{\vdash a,a^{\perp}}\,ax}{\vdash a\oplus a^{\perp},a^{\perp}}\,\oplus_1}{\vdash a\oplus a^{\perp},a\,\&\,a^{\perp}}\,\& \quad \dfrac{\dfrac{\dfrac{\dfrac{}{\vdash b^{\perp},b}\,ax}{\vdash b\oplus b^{\perp},b}\,\oplus_2 \quad \dfrac{\dfrac{}{\vdash b,b^{\perp}}\,ax}{\vdash b\oplus b^{\perp},b^{\perp}}\,\oplus_1}{\vdash b\oplus b^{\perp},b\,\&\,b^{\perp}}\,\& \quad \dfrac{\dfrac{\dfrac{}{\vdash a^{\perp},a}\,ax}{\vdash a\oplus a^{\perp},a}\,\oplus_2 \quad \dfrac{\dfrac{}{\vdash a,a^{\perp}}\,ax}{\vdash a\oplus a^{\perp},a^{\perp}}\,\oplus_1}{\vdash a\oplus a^{\perp},a\,\&\,a^{\perp}}\,\&}{\vdash \beta_B, a\oplus a^{\perp}, b\,\&\,b^{\perp}}\,\beta_B}{\vdash \beta_B, \beta_C, G, (a\oplus a^{\perp}), (a\oplus a^{\perp})}\,\beta_C}{\vdash \beta_A, \beta_B, \beta_C, (G = c\,\&\,c^{\perp})}\,\beta_A$$

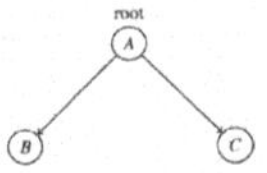

Example 3 Observe the need to use the Mix rule to encode (some) Bayesian networks, as shown by the following example:

$$\mapsto \quad \dfrac{\dfrac{\dfrac{\begin{array}{c}\overline{\vdash a,a^{\perp}}\ \ \overline{\vdash a,a^{\perp}}\\ \vdots \\ \vdash a\oplus a^{\perp}, a\,\&\,a^{\perp}\end{array} \quad \begin{array}{c}\overline{\vdash b,b^{\perp}}\ \ \overline{\vdash b,b^{\perp}}\\ \vdots \\ \vdash b\oplus b^{\perp}, b\,\&\,b^{\perp}\end{array}}{\vdash \beta_B, (a\oplus a^{\perp}), (b\,\&\,b^{\perp})}\,\beta_B \quad \dfrac{\begin{array}{c}\overline{\vdash a,a^{\perp}}\ \ \overline{\vdash a,a^{\perp}}\\ \vdots \\ \vdash a\oplus a^{\perp}, a\,\&\,a^{\perp}\end{array} \quad \begin{array}{c}\overline{\vdash c,c^{\perp}}\ \ \overline{\vdash c,c^{\perp}}\\ \vdots \\ \vdash c^{\perp}\oplus c, c\,\&\,c^{\perp}\end{array}}{\vdash \beta_C, (a\oplus a^{\perp}), (c\,\&\,c^{\perp})}\,\beta_C}{\vdash \beta_B, \beta_C, (a\oplus a^{\perp}), (a\oplus a^{\perp}), b\,\&\,b^{\perp}, c\,\&\,c^{\perp}}\,Mix}{\vdash \beta_A, \beta_B, \beta_C, (G_1 = (b\,\&\,b^{\perp})), (G_2 = (c\,\&\,c^{\perp}))}\,\beta_A$$

where

- the root node A is encoded by the Bayesian fact $\beta_A = (a \oplus a^{\perp}) \parr (a \oplus a^{\perp})$
- the leaf node B is encoded by the Bayesian method $\beta_B = (b \oplus b^{\perp}) \parr (a\,\&\,a^{\perp})$
- the leaf node C is encoded by the Bayesian method $\beta_C = (c \oplus c^{\perp}) \parr (a\,\&\,a^{\perp})$.

Note also in the proof the need for the set of goals $\mathcal{G} = \{G_1 = (b\,\&\,b^{\perp}), G_2 = (c\,\&\,c^{\perp})\}$ which correspond to the terminal links in the proof net representation.

Example 4 The BPN below encodes the BN of Fig. 2, Sect. 3.2: the set of random variables $\{LO, FO, DO, HB, BP\}$ is encoded by the corresponding set of non-terminal B-links, $\{\beta_{LO}, \beta_{FO}, \beta_{DO}, \beta_{HB}, \beta_{BP}\}$ as follows (terminal links have no correspondence in BN):

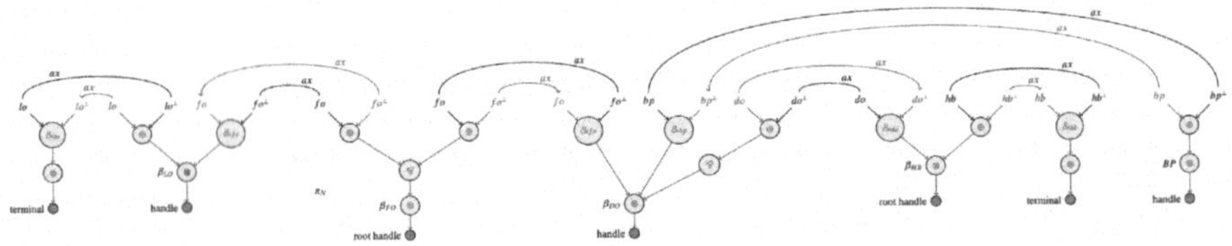

$LO \mapsto \beta_{LO} = (lo \oplus lo^{\perp}) \otimes (fo\,\&\,fo^{\perp})$	(4 switchings)	$FO \mapsto \beta_{FO} = (fo \oplus fo^{\perp}) \parr (fo \oplus fo^{\perp})$	(2 switchings)
$DO \mapsto \beta_{DO} = (do \oplus do^{\perp}) \otimes ((fo\,\&\,fo^{\perp}) \otimes (bp\,\&\,bp^{\perp}))$	(8 switchings)	$HB \mapsto \beta_{HB} = (hb \oplus hb^{\perp}) \otimes (do\,\&\,do^{\perp})$	(4 switchings)
$BP \mapsto \beta_{BP} = (bp \oplus bp^{\perp})$	(2 switchings)		
terminal $T_1 = (lo\,\&\,lo^{\perp})$	(2 switchings)	terminal $T_2 = (hb\,\&\,hb^{\perp})$	(2 switchings)

The other equivalent way to encode "sequentially" the BN of Fig. 2 is represented by the following sequential proof (note the use of the Mix rule):

[16] The right label of the inference indicates the applied method (possibly, a block of inference rules).

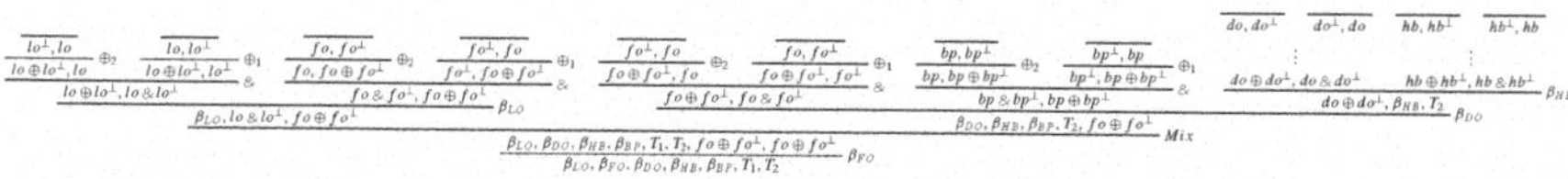

The CPT of every A_i is mapped into the set of restricted slices of the corresponding B-link β_i in a consistent way: each table row is associated to the restricted slice $S(\beta_i)$ that is consistent w.r.t. the input and the output values of the table row. Remind that every random variable A_i is mapped in to a non-terminal B-link β_i whose input (resp., output) interface encodes (possibly, multiple copies of) the boolean values "True" and "False" for the variables that are parents of A_i (resp. for the variable A_i itself): as seen before, each instance of a positive literal encodes the value "True" while each instance of a negative literal encodes the value "False". Observe that, excluding the terminal links, we have overall 20 additive slices (conditional probability values) for the B-links against $2^5 = 64$ possible slices (joint probabilities) of π. This is consequence of the fact that each slice for a proof net π can be viewed as a product (a composition) of the partial slices of the B-links of π; therefore, to calculate the joint probability of a slice we can exploit the slice decomposition that is the analogue of the factorization (via chain rule) of the joint probability. Example: to infer the the joint probability $Pr(\neg lo, fo, \neg do, hb, \neg bp)$, select the switching $S_k(\pi^{\mathcal{N}})$, corresponding to the grey sub-graph of the previous BPN, which retains the premises $lo^\perp, fo, do^\perp, hb, bp^\perp$ for, resp., the links $\&_{lo}$, $\&_{fo}$, $\&_{do}$, $\&_{hb}$ and $\&_{bp}$ (delete, resp., their dual premises together with the axiom links that have these duals as conclusions). Then, by slice decomposition, we can compute the probability

$$\begin{aligned}\langle S_k(\pi^{\mathcal{N}})\rangle &= \langle S_k(\beta_{FO}\cdot\beta_{BP}\cdot\beta_{DO}\cdot\beta_{LO}\cdot\beta_{HB})\rangle\\ &= \langle S_k(\beta_{FO})\cdot S_k(\beta_{BP})\cdot S_k(\beta_{DO})\cdot S_k(\beta_{LO})\cdot S_k(\beta_{HB})\rangle\\ &= \langle S_k(\beta_{FO})\rangle\cdot\langle S_k(\beta_{BP})\rangle\cdot\langle S_k(\beta_{DO})\rangle\cdot\langle S_k(\beta_{LO})\rangle\cdot\langle S_k(\beta_{HB})\rangle\end{aligned}$$

corresponding to the factorization of $Pr(\neg lo, fo, \neg do, hb, \neg bp)$ that is:

$$Pr(fo)Pr(\neg bp)Pr(\neg do|fo, \neg bp)Pr(\neg lo|fo)Pr(hb|\neg do).$$

Finally, the following sub-proof net of $\pi^{\mathcal{N}}$ returns the marginalization $Pr(lo, do)$ optimized via "sums shifting" and "irrelevant variable elimination" of Eq. (6) in Sect. 3.2: the dotted l.h.s. (resp. r.h.s.) factor corresponds to a graph that is the superposition of two slices depending on β_{FO} (resp., 2^2 slices over β_{BP} and β_{DO}). Note that the B-link β_{HB} is irrelevant, so omitted.

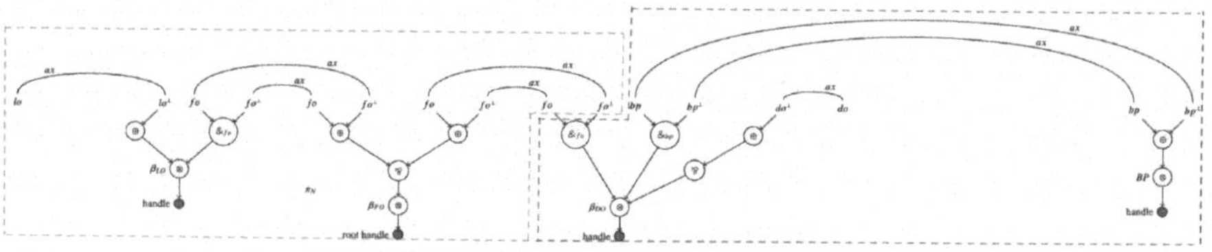

$Pr(lo, do) = \sum_{FO} Pr(FO)Pr(lo|FO) \sum_{BP} Pr(BP)Pr(do|FO, BP)$

5 Conclusions and Further Works

We proposed a notion of *probabilistic (or Bayesian) proof* for the MALL fragment of linear logic, both in graphical (proof-nets) and sequential style (derivations), which is correct and complete with respect to the notion of Bayesian network. Analogously to what happens with Bayesian networks, Bayesian proofs allow both an intelligible and efficient reasoning on probabilities such, e.g., marginalization of probabilities, etc.

Finally, concerning possible future applications, we mention the implementation of a *probabilistic logic programming paradigm* based on probabilistic (Bayesian) proofs in which the mechanism of probabilities reasoning (illustrated in Sect. 4.5.4) is implemented as *sequent proof search* or *proof-net construction* [3, 5].

Acknowledgements Research supported by the Istituto Nazionale di Alta Matematica "Francesco Severi" (INdAM, GNSAGA Research Group).
The author thanks the anonymous referees, Matteo Acclavio and Luca Roversi for the for the useful discussions on the topic of proof nets and probability distributions.

References

1. Abrusci VM, Maieli R (2016) Cyclic multiplicative-additive proof nets of linear logic with an application to language parsing. In: Foret A, Morrill G, Muskens R, Osswald R, Pogodalla S (eds) Formal grammar—20th and 21st international conferences, FG 2015, Barcelona, Spain, August 2015, Revised Selected Papers. FG 2016, Bozen, Italy, August 2016, Proceedings, Lecture Notes in Computer Science, vol 9804. Springer, pp 43–59 (2015). https://doi.org/10.1007/978-3-662-53042-9_3
2. Abrusci VM, Maieli R (2015) Cyclic multiplicative proof nets of linear logic with an application to language parsing. In: de Paiva V, de Queiroz R, Moss LS, Leivant D, de Oliveira AG (eds) Logic, language, information, and computation. Springer, Berlin, Heidelberg, pp 53–68. https://doi.org/10.1007/978-3-662-47709-0_5
3. Abrusci VM, Maieli R (2019) Proof nets for multiplicative cyclic linear logic and Lambek calculus. Math Struct Comput Sci 29(6):733–762. https://doi.org/10.1017/S0960129518000300
4. Acclavio M, Maieli R (2020) Generalized connectives for multiplicative linear logic. In: Fernández M, Muscholl A (eds) 28th EACSL annual conference on computer science logic, CSL 2020, 13–16 Jan 2020, Barcelona, Spain, LIPIcs, vol 152. Schloss Dagstuhl - Leibniz-Zentrum für Informatik, pp 6:1–6:16. https://doi.org/10.4230/LIPICS.CSL.2020.6
5. Acclavio M, Maieli R (2024) Logic programming with multiplicative structures. Electron Proc Theor Comput Sci 408:42–61. https://doi.org/10.4204/EPTCS.408.3
6. Charniak E (1991) Bayesian networks without tears. AI Mag 12(4):50
7. Cooper GF (1990) The computational complexity of probabilistic inference using Bayesian belief networks. Artif Intell 42(2–3):393–405. https://doi.org/10.1016/0004-3702(90)90060-d
8. Danos V (1990) Une application de la logique linéaire a l'etude des processus de normalisation (principalement du λ-calcul). Thèse de doctorat, Université Paris VII, Paris, France. https://perso.ens-lyon.fr/pierre.lescanne/PUBLICATIONS/DanosPhD.pdf
9. Danos V, Regnier L (1989) The structure of multiplicatives. Arch Math Logic 28(3):181–203. https://doi.org/10.1007/bf01622878

10. Darwiche PA (2009) Modeling and reasoning with Bayesian networks, 1st edn. Cambridge University Press, USA. https://doi.org/10.5555/1534901
11. Gentzen G (1969) The collected papers of Gerhard Gentzen. North-Holland Pub. Co., Amsterdam
12. Girard JY (1987) Linear logic. Theoret Comput Sci 50(1):1–101
13. Kogkalidis K, Moortgat M, Moot R (2020) Neural proof nets. In: Fernández R, Linzen T (eds) Proceedings of the 24th conference on computational natural language learning. Association for Computational Linguistics, Online, pp 26–40. https://doi.org/10.18653/v1/2020.conll-1.3. https://aclanthology.org/2020.conll-1.3/
14. Laurent O, Maieli R (2008) Cut elimination for monomial MALL proof nets. In: Proceedings of the twenty-third annual IEEE symposium on logic in computer science, LICS 2008, 24–27 June 2008, Pittsburgh, PA, USA. IEEE Computer Society, pp 486–497. https://doi.org/10.1109/LICS.2008.31
15. Maieli R (2007) Retractile proof nets of the purely multiplicative and additive fragment of linear logic. In: Dershowitz N, Voronkov A (eds) Logic for programming, artificial intelligence, and reasoning, 14th international conference, LPAR 2007, Yerevan, Armenia, 15–19 Oct 2007, Proceedings, Lecture Notes in Computer Science, vol 4790. Springer, pp 363–377. https://doi.org/10.1007/978-3-540-75560-9_27
16. Maieli R (2014) Construction of retractile proof structures. In: Dowek G (ed) Rewriting and Typed Lambda Calculi—joint international conference, RTA-TLCA 2014, held as part of the Vienna summer of logic, VSL 2014, Vienna, Austria, 14–17 July 2014. Proceedings, Lecture Notes in Computer Science, vol 8560. Springer, pp 319–333. https://doi.org/10.1007/978-3-319-08918-8_22
17. Maieli R (2019) Non decomposable connectives of linear logic. Ann Pure Appl Log 170(11). https://doi.org/10.1016/j.apal.2019.05.006
18. Maieli R (2022) A proof of the focusing theorem via MALL proof nets. In: Ciabattoni A, Pimentel E, de Queiroz RJGB (eds) Logic, language, information, and computation—28th international workshop, WoLLIC 2022, Iaşi, Romania, 20–23 Sept 2022, Proceedings, Lecture Notes in Computer Science, vol 13468. Springer, pp 1–17. https://doi.org/10.1007/978-3-031-15298-6_1
19. Maieli R (2025) Bayesian proof nets. In: Mehmood R, Hernández G, Praça I, Wikarek J, Loukanova R, Monteiro dos Reis A, Skarmeta A, Lombardi E (eds) Distributed computing and artificial intelligence, special sessions I, 21st international conference. Springer Nature Switzerland, Cham, pp 142–152. https://doi.org/10.1007/978-3-031-76459-2_13
20. Pearl J (2009) Causality, 2nd edn. Cambridge University Press
21. Raza S, Haider S (2010) Modeling first-order Bayesian networks (FOBN). In: 2010 3rd international conference on advanced computer theory and engineering (ICACTE), vol 2, pp V2-571–V2-575. https://doi.org/10.1109/ICACTE.2010.5579472
22. Soldát M, Kléma J (2024) Directed evolution of proteins via Bayesian optimization in embedding space. In: Cannataro M, Zheng HJ, Gao L, Cheng J, de Miranda JL, Zumpano E, Hu X, Cho Y, Park T (eds) IEEE international conference on bioinformatics and biomedicine, BIBM 2024, Lisbon, Portugal, 3–6 Dec 2024. IEEE, pp 91–98. https://doi.org/10.1109/BIBM62325.2024.10822356

Skolem Functions for Russian Negative Concord: Developing Preprocessing Algorithm

Ulyana Zaitseva and Alexey Stukachev

Abstract In linguistics and mathematics, the phenomenon of negative concord presents a challenge for analysis using the methods of formal semantics, particularly when relying on the principle of compositionality. For the Russian language, we propose a potential solution for formalizing sentences with negative concord. This solution involves a linguistic pre-processing of the sentences, followed by the application of a mathematical tool: Skolem functions. This paper presents an algorithm for the linguistic pre-processing of Russian sentences with canonical word order (subject-verb-object) and discusses how Skolem functions can be applied to formally represent negative concord. This approach allows for a more precise analysis and formalization of the semantic structure of certain linguistic phenomena in the Russian language.

Keywords Mathematical linguistics · Skolem functions · Generalized quantifiers · Negative concord · Negative indefinites

1 Introduction

This paper continues previous research [1] comparing the methods of generalized quantifiers and Skolem functions in the analysis of negative concord. *Negative Concord or NC* refers to a grammatical phenomenon characterized by the co-occurrence of several formally negative elements within a sentence to express a single semantic negation [2].

U. Zaitseva (✉) · A. Stukachev
Novosibirsk State University,Novosibirsk, Russia
e-mail: u.penzina@g.nsu.ru

A. Stukachev
e-mail: aistu@math.nsc.ru

A. Stukachev
Sobolev Institute of Mathematics,Novosibirsk, Russia

R. Loukanova (ed.), *Computational Linguistics, Information, Reasoning, and AI 2024*, Studies in Computational Intelligence 1280,
https://doi.org/10.1007/978-3-032-21143-9_6

The interpretation of Russian negative pronouns such as *nikto, nichego, ni odin* (eng. nobody, nothing, no one) is difficult because those words are often seen as having their own negative meaning. This creates problems for the principle of compositionality, which states that the meaning of a sentence comes from the combined meaning of its parts.

In techniques involving generalized quantifiers, attempts have been made to construct a formal representation for such sentences [3, 4]. These efforts were based on the idea that negative pronouns could be viewed as generalized quantifiers (a quantifier expression refers to a set of sets of individuals [5]. From the perspective of types, a generalized quantifier has a functional type $(e \to t) \to t)$. However, this approach complicates the original meaning of the sentence.

In this paper, we adopt the perspective [6, 7] according to which negative concord is a phenomenon of syntactic agreement. This agreement occurs between a negative operator with interpretable negative features (often a sentential negation marker, e.g., "it's not true that") and one or more negative agreement elements that carry uninterpretable negative features. The negative pronouns found in such sentences don't actually carry a negative meaning themselves. Instead, they're treated as indefinite words that are simply placed under the scope of a negative operator [8].

Given the presence of a negative concord marker and an existential component in the elements under negation, we propose to treat them as Skolem choice functions. As we will discuss below, these functions are a convenient and useful tool for analyzing sentences with an existential structure. They have a simpler functional type that is not tied to a truth or falsehood value $(e \to t) \to e$.

This paper thus offers a novel approach to addressing these challenges, involving the preprocessing of linguistic material prior to direct analysis. We hope to contribute to a more accurate and comprehensive understanding of negative concord items in linguistic structures.

This paper is organized into four main sections. In Sect. 2 we define key terms and distinguish between Negative Polarity Items and Negative Concord Items. In Sect. 3 we then compare the use of generalized quantifiers with Skolem functions to show why the latter is a more suitable tool for our analysis. Following this, in Sect. 4 we present a step-by-step algorithm for pre-processing Russian sentences with negative concord, applying it to both English and Russian examples to demonstrate its effectiveness. Finally, in Sect. 5 we conclude with a summary of our findings and a brief discussion of the broader implications for the study of natural language semantics.

2 Negative Concord Items

2.1 Negative Concord

Researchers [9] often identify two classes of negative concord languages: strict and non-strict. Languages with non-strict negative concord, such as Italian and Spanish, are those in which postverbal negative indefinites participate in negative concord when combined with a sentential negation marker. In contrast, languages with strict NC, like Russian, Polish, and Hungarian, are those in which a negative indefinite pronoun (such as *nobody* or *nothing*) must co-occur with a sentential negation marker (such as *not* or *ne*) [10]. This is a rule of grammar, not an optional choice for emphasis. The term "concord" is employed here in an extended sense, transcending its conventional application to attribute-noun agreement patterns, such as adjective-noun congruence. Within English linguistic scholarship, the designation "negative concord" is preferentially utilized over alternative phrasings like "negative agreement" [11].

As outlined in a recent study [12], the main idea behind Negative Concord is that a single semantic negation can show up in two places at once: both as part of the verb and in negative adverbs, pronouns, or determiners. Take the Russian phrase, "*Nikto nichego ne znaet*" (Nobody knows anything) for example. Even though it has three negative elements—*nikto*, *nichego*, and *ne*—it expresses just one, simple negation. The types of negative concord operators that fall into this category are often called "n-words". These include:

- Russian: *nikto*, *nichto*, *nikakoy*, *ni odin*.
- English: *nobody*, *nowhere*, *nothing*.

This phenomenon highlights the diverse ways languages handle negation. It showcases a clear typological divide between languages that use negative concord (where multiple negative words express a single negation) and those that enforce a prohibition against double negation, typically seen in Standard English.

2.2 Negative Polarity Items Versus Negative Concord Items

Cross-linguistic studies reveal significant diversity in the typology of Negative Polarity Items (NPIs), which are words that can only appear in certain contexts. English, for instance, has a notable distinction between *weak* and *strong* NPIs [13].

Weak NPIs, such as *anyone*, *any*-NPs, and *ever*, are permitted in contexts that are "non-veridical", meaning they don't necessarily express a truth. These include questions, conditionals, and comparatives [14]. In contrast, strong NPIs—including negative pronouns like *no one*, *nothing*, *nowhere*—are restricted to overtly negative environments. While English NPIs are subject to these specific licensing conditions, a different phenomenon is observed in languages with negative concord. Typically, negative indefinites like Russian *nikto* ("nobody") are considered negative concord items (hereinafter—NCIs). This means they are assigned a non-negative meaning and carry uninterpretable negative features. NCIs are words or phrases that must appear with a negative marker to form a single negation [10].

To explain the distribution of NPIs, researchers have developed various semantic theories [15, 16]. The goal of these theories is to identify the specific properties of a linguistic environment that allow an NPI to appear. These properties are referred to as licensing conditions, and the element that creates such a context is called a licenser. The debate among linguists centers on what exactly constitutes a valid licenser for different types of NPIs.

The initial theory for why NPIs are allowed in certain places, proposed by Ladusaw [16], suggested that they are only acceptable within the scope of *monotonically decreasing functions*. However, later research has expanded this idea to cover a wider range of contexts. For example, Giannakidou [15] showed that weak NPIs are allowed based on the *non-veridicality* of their surrounding context, not just strict downward monotonicity.

Definition 1 (Veridical Context (Preliminary)) A context C[] is *veridical* if $C[p]$ implies p. [13]

This definition was later revised to emphasize entailment relations: a context C[] is veridical iff inserting a proposition-denoting expression p into C[] yields a meaning that entails p.

Recent typological studies have provided a clearer picture of strong NPIs, especially in languages with negative concord. For instance, research on Russian by Refs. [17, 18] has identified a specific group of negative concord items (NCIs) that must appear with a sentential negation marker. This shows that the rules for where NPIs can appear vary from one language to another, depending on both syntax (word order) and semantics (meaning).

3 Generalized Quantifiers and Scolem Functions for Negative Concord

When we study negative concord in the formal semantics of natural language, it requires specific methods to account for how negative elements interact. This section compares two approaches: generalized quantifiers and Skolem functions. Both offer ways to formalize the semantics of negative constructions, but they model their structure differently.

3.1 Generalized Quantifiers

The theory of generalized quantifiers, developed in the works [19–21] provides a flexible way for analyzing quantificational expressions. Within this framework, quantifiers are interpreted as functions that take predicates and return truth values (true or false). For example, the quantifier "somebody" (someone) can be represented as a function of type $(e \rightarrow t) \rightarrow t$, where e is the type of entities and t is the type of truth values. Formally, this is expressed as follows:

$$[\text{somebody}] = \lambda P.\exists x.(\text{Person}(x) \wedge P(x)).$$

This approach allows for the analysis of statements such as "Some student buys coffee" through the combination of a quantifier and a predicate.

However, this approach runs into problems with languages that use negative concord, such as Russian. In these languages, multiple negative words (*nikakoy* (no) and *ne* (not)) are used to express a single negation. A direct application of generalized quantifiers here can lead to a semantic contradiction because the formal interpretation doesn't account for the grammatical link between these negative parts.

3.2 Skolem Functions

Skolem functions, tool from mathematical logic, introduced by Thoralf Albert Skolem in the early twentieth century, help us deal with quantification in language. They work by assigning a specific value to a variable in a sentence, which helps us to understand the meaning of phrases like "someone" or "something". Essentially, a Skolem function represents a choice -it picks out a specific "witness" or individual from a set. This ability to "choose" a representative makes the existence of that individual explicit, transforming a statement about an unknown "someone" into a statement about a specific, named (though still placeholder) individual.

This method is particularly useful for analyzing sentences with multiple quantifiers or indefinite noun phrases, as demonstrated in Robaldo [22] framework for analyzing sentences with nested quantifiers and Winter [23] investigations into the interpretation of indefinite nominal phrases. Within this theoretical framework, choice functions resolve scope ambiguities by algorithmically selecting witnesses for existential quantifiers, thereby mapping syntactic structures to precise semantic representations. Further extending this approach, Yanovich [24] employs Skolem functions to analyze Russian indefinite pronouns marked by *-to* and *-nibud'*.

Skolem functions represent a formal method for eliminating existential quantifiers. The procedure of skolemization is based on replacing variables bound by existential quantifiers (∃) with constants or functional symbols, allowing formulas to be transformed into equivalent forms without existential quantifiers. This approach plays a crucial role in automated theorem proving and the formal analysis of natural language [25].

Definition 2 Let φ be a formula of the form:

$$\varphi = \forall y_1 \ldots \forall y_n \exists x \, \psi(x, y_1, \ldots, y_n),$$

where ψ is a formula that does not contain existential quantifiers. A **Skolem function** f is introduced to replace the variable x, which depends on the universally quantified variables $y_1, \ldots, y_n$:

$$\varphi_{sk} = \forall y_1 \ldots \forall y_n \, \psi(f(y_1, \ldots, y_n), y_1, \ldots, y_n).$$

If $n = 0$ (i.e., the existential quantifier does not depend on any universal quantifiers), then f becomes a constant c:

$$\varphi_{sk} = \psi(c).$$

Example 1

Original Formula: $\forall y \exists x$ Love (y, x) (reads as "for every person y, there exists a person x that y loves")

After Skolemization: $\forall y$ Love $(y, f(y))$, when we apply Skolemization, we replace the existential quantifier's variable x with a Skolem function $f(y)$. The new formula means "for every person y, y loves the person that the function f selects". This function is a placeholder for the specific person x that exists for each y. In essence, it captures the idea that the existence of x depends on the choice of y.

Example 2

Original Formula: $\exists x$ Swim (x) (reads as "there exists some x that can swim")

After Skolemization: Swim(c), in this case, the existential quantifier is not dependent on any universal quantifiers. Therefore, after Skolemization, we replace the variable x with a Skolem constant (c). The new formula, Swim(c), now means "the specific person c can swim". The constant (c) is a placeholder for the individual who can swim, affirming their existence without the use of an existential quantifier.

4 Building the Algorithm

Before we formalize the analysis, a few key terms need to be defined:

Definition 3 A **negative pronoun** is a non-referential indefinite pronoun ending in *-nibud'*, which, when combined with negation, falls within its scope. For example, *"nikto"* is semantically equivalent to (*neverno, chto + kto-nibud'*) [17].

Definition 4 Cumulative negation is a type of negation that attaches to a pronoun ending in *-nibud'* or *hot' odin* (which express existential quantification in Russian). It forms a negative ni-pronoun but requires that the sentential negation marker (ne) is not attached to the verb. [11].

Using these definitions, we can examine how to transform Russian sentences with negative concord. The goal is to prepare them for analysis with Skolem functions. We assume that a transformed sentence—where a sentential negation combines with an indefinite pronoun—accurately represents the original sentence with negative concord.

The procedure for our analysis has three steps:

1. **Identify Negative Concord**: Check if the sentence has both a negative pronoun and a negative particle on the verb.
2. **Transform the Pronoun**: Change the negative pronoun into its non-referential indefinite equivalent. Introduce a sentential negation marker (e.g., *"it is not the case that"* at the beginning of the sentence).
3. **Remove the Cumulative Negation**: Take away particle (the particle *"ne"* associated with the verb, if present).

After these steps, we can construct the logical representation of the sentence with negative concord. This involves using a Skolem function to formalize the existence of an object and building a syntactic parse tree.

4.1 Analyzing the English Sentence

Let's first apply this algorithm to the English sentence, *"No one writes to him"* This will clearly show why our pre-processing steps are necessary only for languages like Russian.

1. **Identify Negative Concord**: This sentence has a negative pronoun, "no one", but lacks a second negative particle on the verb. Therefore, it does not exhibit negative concord.
2. **Transform the Negative Pronoun**: In this case, "no one" is already a non-referential negative pronoun.
3. **Remove the Cumulative Negation Particle**: This step is also not necessary as there is no cumulative negation particle in the sentence.

The logical representation of the English sentence, therefore, does not require our algorithm's pre-processing. The sentence can be directly formalized using a generalized quantifier, with "no one" acting as a function that takes the predicate "writes to him" and returns a truth value. The logical representation of the sentence is:

$$(\lambda P.\neg\exists x.(\mathrm{Person}(x) \wedge P(x))(\lambda x.\mathrm{Write}(x, \mathrm{him})))$$

This simplifies to:

$$\neg\exists x.(\mathrm{Person}(x) \wedge \mathrm{Write}(x, \mathrm{him}))$$

The logical representation correctly reflects the original meaning. This sentence can be represented with a syntactic parse tree based on generalized quantifiers, as shown below:

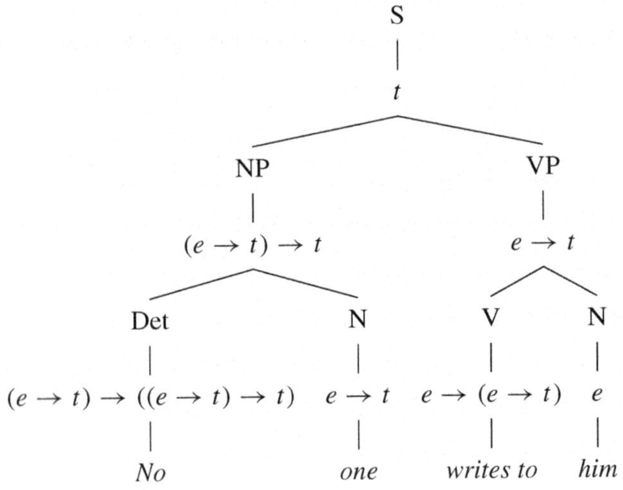

4.2 Analyzing the Russian Sentence

Now, let's turn to the Russian sentence, *"Nikakoj chelovek ne pishet emu"*, which poses a significant challenge for the principle of compositionality if we don't pre-process it.

A direct, word-for-word analysis would incorrectly treat *nikakoj chelovek* ("no person") as a negative quantifier and *ne pishet* ("does not write") as a second negation. This leads to a complex and ultimately incorrect formal representation, as the two negatives would cancel each other out:

$$(\lambda P.\neg\exists x.(\mathrm{Person}(x) \wedge \neg P(x))(\lambda x.\mathrm{Write}(x, \mathrm{h})))$$

This expression, when simplified, leads to a contradictory and incorrect meaning. To avoid this, we must apply our pre-processing algorithm.

1. **Identify Negative Concord**: *"Nikakoj chelovek ne pishet emu"*—the sentence contains the negative pronoun *nikakoj* and the negative particle *ne* with the verb, indicating negative concord.
2. **Transform the Negative Pronoun**: We change the negative pronoun *nikakoj* to its non-referential indefinite equivalent, *kakoj-nibud'*. This results in the sentence: *"Kakoj-nibud' chelovek ne pishet emu"*.
3. **Remove the Cumulative Negation Particle**: We remove the *ne* particle associated with the verb, which gives us: *"Neverno, chto kakoj-nibud' chelovek pishet emu"*.

By following these steps, we arrive at a sentence that is semantically equivalent to the original but is now compatible with a logical framework. We can now formalize this pre-processed sentence:

$$(\lambda P.\neg\exists x.(\mathrm{Person}(x) \wedge P(x))(\lambda x.\mathrm{Write}(x, \mathrm{h})))$$

$$\neg\exists x(\mathrm{Person}(x) \wedge \mathrm{Write}(x, \mathrm{h}))$$

This representation precisely matches the logical form of the English sentence. We can then introduce a Skolem constant, c, to eliminate the existential quantifier, resulting in:

$$\neg(\mathrm{Person}(c) \wedge \mathrm{Write}(c, \mathrm{h}))$$

This revised, correct logical representation accurately captures the original meaning. Furthermore, we can now construct a valid syntactic parse tree for the pre-processed sentence.

However, the syntactic parse tree for the sentence *"Nikakoj chelovek ne pishet emu"* using functional types of generalized quantifiers cannot be constructed. This is because the representation that *"nikakoj chelovek"* is a negative quantifier requires the truth value of the statement to change twice, thereby making the sentence equivalent to *"Kazhdyj chelovek pishet emu"*, which does not correspond to reality. Therefore, we will consider the parse tree of the sentence *"Neverno, chto kakoj-nibud' chelovek pishet emu"* using functional types of Skolem functions:

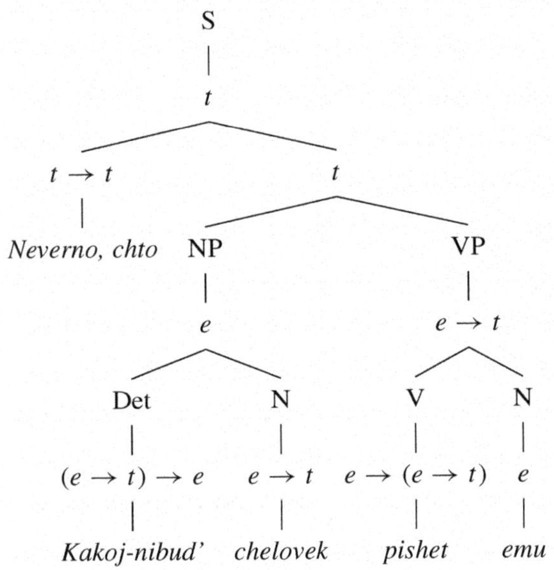

This analysis demonstrates the critical role of our linguistic pre-processing algorithm in ensuring that the formal representation of sentences with negative concord is both logically valid and semantically faithful to the original meaning. After pre-processing, noun phrases with indefinite pronouns can be analyzed using generalized quantifiers, but choice functions have a simpler functional type and more accurately reflect the semantics of natural language expressions.

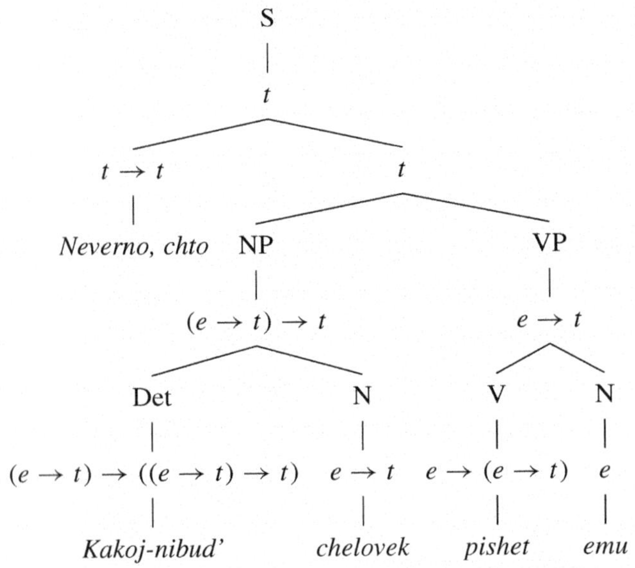

4.3 List of Examples

The sentences listed below,[1] serve as a foundational set of examples for our algorithm's linguistic pre-processing. Specifically, they all follow a simple Subject-Verb-Object (SVO) word order and exhibit strict negative concord.

This initial focus allows us to test our algorithm on the most basic sentence structures, as each example contains one or more negative indefinites (like "nikto" or "nichego") that obligatorily co-occur with a sentential negation marker ("ne"). This is the exact type of phenomenon our algorithm is designed to handle.

Future research will build on this foundation by applying the algorithm to more complex sentence types and nested structures.

1. Nikto ne kontroliruet! [Forum: Univer (institut) VS shkola. Plyusy i minusy. Gde v itoge luchshe i pochemu? (2011)]
2. Nikto ne proveryaet tebya, ne zastavlyaet uchit'sya, kak eto bylo v shkole [Tam zhe].
3. Nikto ne vspomnit ego dobrym slovom [V. S. Savel'ev. Dnevnik (1974)].
4. Nichyego ne poluchaetsya s kartinoj. Nikto nichego ne ponimaet [A. A. Tarkovskij. Martirolog. Dnevniki (1974)].
5. Nichyto ne ischezayet bez sleda [Sever Gansovskij. Mlechnyj put' // "Tekhnika - molodezhi", 1974].
6. Nichyto v mire ne zamenit samogo poslednego vuzovskogo atraktsiona: sdachi diplom [Forum: Shkola ili univer gde legche?)) (2006)].
7. Nichyto tak ne podnimaet v glazakh obyvatelya prestizh nauki, kak teorii o kontse sveta [Sergej Leskov. Ne spasutsya dazhe cherti. Sozdana eshche odna teoriya o kontse sveta // "Izvestiya", 2002.06.14].
8. Nikakie ryzhie ko mne ne prikhodili [Arkadij Strugatskij, Boris Strugatskij. Za milliard let do kontsa sveta (1974)].
9. Nikakaya gladil'naya doska tut ne nuzhna [Aleksandr Volkov. Odetsya s igolochki i bez nitochki // "Znanie – sila", 2003].
10. Ya ponjala: tebya nikakaya devushka ne zhdet, tak ya budu zhdat',—i, utirayas' ladoshkoj, dobavila:—Pravda-pravda! [Viktor Astaf'ev. Oberton (1995–1996)].
11. Nikakoj ya ne borets,—skazal ya, uzhe nyryaya pod odeyalo,—razdevaytes' i gasite svet [Fazil' Iskander. Muki sovesti, ili Bajskaya krovat' (1980–1990)].
12. Nikakoe, odnako, ustrojstvo ne pryatalos' v tesnoj komnate Bashutskogo M. A.: stepen' ego ispuga prebyvala vpolne dostatochnoj dlya obespecheniya sotsial'noj stabil'nosti [Yurij Davydov. Sinie tyul'pany (1988–1989)].
13. Nikakoj on ne genij, on ustarelyj krikun, pugach, fanfaron...[Danijl Granin. Zubr (1987)]
14. Nikakaya epidemiya stol'ko ne kosit, kak eto zimnee bedstvie. I. Grekova. Perelom (1987)].

[1] Here and further, examples are taken from the National Corpus of the Russian Language (ruscorpora.ru).

15. Nikakaya opredelenaya matematicheskaya skhema ne ischerpyvaet vsej konkretnosti deystvitel'nykh yavlenij...[A. N. Kolmogorov. Matematika (1954–1987)]
16. Sovremennaya matematicheskaya logika dala na etot vopros opredelennyj otvet: nikakaya edinaya deduktivnaya teoriya ne mozhet ischerpat' raznoobraziya problem teorii chisel [A. N. Kolmogorov. Matematika (1954–1987)].
17. Nikakaya golova s etim ne spravitsya [Anatolij Rybakov. Tyazhelyj pesok (1975–1977)].
18. Nikakoe vneshnee bytie ne mozhet ego opredelit'! [Aleksandr Solzhenitsyn. V kruge pervom, t.1, gl. 26–51 (1968) // "Novyj Mir", 1990].
19. Nikakaya privychka k tyazheloj rabote ne menyala tut rovno nichego [Varlam Shalamov. Kolymskie rasskazy (1954–1961)].
20. Nikakaya teologiya ne reguliruet vneshne moyego protsessa poznaniya i ne stavit pered menya norm [N. A. Berdyaev. Novoe srednevekov'e (1924)].

5 Conclusions

In this paper, we tackled the challenge of formally representing Russian sentences that use negative concord. Our approach offers a new perspective on how to analyze these linguistic phenomena without sacrificing the principle of compositionality.

The core principle of our proposed solution is the use of Skolem functions. We suggest that by treating negative pronouns like *nikto* ("nobody') and *nichto* ("nothing') not as intrinsically negative, but as non-negative indefinite pronouns under negation, we might simplify their formal representation. This approach seeks to avoid the semantic conflicts that have complicated earlier theories.

This method, however, is applied with a specific limitation in mind. It relies on a linguistic pre-processing algorithm that is, at this stage, designed to work only with simple sentence structures. This algorithm prepares the Russian sentence for formal analysis by systematically transforming the negative pronouns and accurately distinguishing them from the sentential negation marker. While the effectiveness of this preliminary approach is demonstrated through our analysis of several examples, we acknowledge that its applicability is currently restricted to less complex grammatical constructions.

This application of Skolem functions thus presents a viable framework for addressing the complexities of negative concord.

Acknowledgements The second author was supported by the IM SB RAS state agreement, project number FWNF-2022-012.

References

1. Penzina U, Stukachev A (2025) Quantifiers for negative polarity items. In: Distributed computing and artificial intelligence, Special Sessions I, 21st International Conference, vol 1198. Springer Nature, p 123
2. Corblin F, Déprez V, De Swart H, Tovena L (2004) Negative concord. Handbook of French semantics, pp 417–452
3. Coppock E, Champollion L (2019) Invitation to formal semantics. Boston University and New York University. eecoppock.info/semantics-boot-camp.pdf
4. Peters S, Westerståhl D (2006) Quantifiers in language and logic. OUP Oxford
5. Van Benthem J (1984) Questions about quantifiers1. J Symb Log 49(2):443–466
6. Penka D (2001) Negative indefinites. Oxford University Press, p 32
7. Zeijlstra H (2004) Sentential negation and negative concord. Ph.D. thesis, External Organizations (2004)
8. Rossyaykin P (2022) Negative concord, fragments and (downward) agree. Paper presented at WCCFL 40
9. Giannakidou A (2006) N-words and negative concord. The Blackwell companion to syntax, pp 327–391
10. Jeretič P (2025) Is negation negative? (And a discussion of negative concord in SOV languages). Languages 10(6):130
11. Paducheva E (2001) Negation. Materials for the project of corpus description of Russian grammar (http://rusgram.ru), on the rights of manuscript
12. Van der Auwera J, Van Alsenoy L (2016) On the typology of negative concord. Stud Lang 40(3):473–512
13. Barker C (2018) Negative polarity as scope marking. Linguist Philos 41(5):483–510
14. Partee BH, Borschev V (2002) Genitive of negation and scope of negation in Russian existential sentences. In: Annual workshop on formal approaches to Slavic linguistics: the second Ann Arbor meeting 2001 (FASL 10), pp 181–200
15. Giannakidou A (1998) Polarity sensitivity as (non) veridical dependency. John Benjamins Publishing Company
16. Ladusaw WA (1979) Polarity sensitivity as inherent scope relations. The University of Texas at Austin
17. Paducheva E (2011) Implicit negation and pronouns with negative polarization. Voprosy Yazykoznaniya 1:3–18
18. Rossyaykin P (2020) Existential modal and negation in Russian: evidence for universal functional hierarchy. In: Proceedings of ConSOLE XXVIII, pp 136–155
19. Barwise J, Cooper R (1981) Generalized quantifiers and natural language. In: Philosophy, language, and artificial intelligence: resources for processing natural language. Springer, Cham, pp 241–301. https://doi.org/10.1007/978-94-009-2727-8_10
20. Keenan EL, Westerståhl D (1997) Generalized quantifiers in linguistics and logic. In: Handbook of logic and language. Elsevier, pp 837–893. https://doi.org/10.1016/B978-044481714-3/50020-5
21. Montague R (1973) The proper treatment of quantification in ordinary English. In: Approaches to natural language: proceedings of the 1970 Stanford workshop on grammar and semantics. Springer, Cham, pp 221–242. https://doi.org/10.1007/978-94-010-2506-5_10
22. Robaldo L (2008) Skolem theory and generalized quantifiers. In: International workshop on logic, language, information, and computation. Springer, Cham, pp 286–297
23. Winter Y (2004) Functional quantification. Res Lang Comput 2:331–363
24. Yanovich I (2005) Choice-functional series of indefinite pronouns and Bamblin semantics. In: Semantics and linguistic theory, pp 309–326
25. Mirrazi Z (2019) Indefinites: an argument for skolemization with world variables. Linguistics in Göttingen: Colloquium Series

Correlations Between Verb Meaning and Verb Valence

Lars Hellan

Abstract Based on a large valence resource where verbs are systematically aligned with meanings and sets of valence frames associated with each verb, we explore whether one can find overall tendencies pertaining to *which types of valence specifications* relate to *which types of meaning*. We discuss examples of possible correlations, stated in descriptive, linguistic terms using the main units and principles of organization of the resource. The resource is a system *NorVal-1952* of Norwegian Hellan [1], a subpart of the valence lexicon *NorVal* (Hellan [2]; Hellan [3]) where a system of 'semantic meta-features' has been added to 1952 verb lexemes (For NorVal-1952, cf. (Hellan L [4]), and for NorVal, cf. (Hellan L [5]; Hellan L [6]). An English counterpart to NorVal-1952 has recently also been built, so that the investigations and their outcomes can possibly be replicated for English, and in turn for other languages.

Keywords Meaning feature · Valence · Meaning-valence correlation · Norwegian

1 Introduction

Is it possible to state general correlations between verb valence frames and verb meanings, such that in a large inventory of verb valence frames, one can predict from the meaning of a verb which valence frame(s) it can take, and from the valence frames that a verb takes, what its meaning will be?

A tradition of studies exploring this kind of question is the one of [7], VerbNet [8, 9]. Here the circumstance of a verb having more than one valence frame is referred to as an instance of a 'valence alternation', building on observations to the effect that variations in valence patterns can highlight parameters of meaning, and that the parameters may be shared across verbs, thus pointing to possible semantic regularities among verbs having multiple frames, leading, e.g., to the notion of 'Verb Classes'.

L. Hellan (✉)
Norwegian University of Science and Technology (NTNU), Trondheim, Norway
e-mail: lars.hellan@ntnu.no

R. Loukanova (ed.), *Computational Linguistics, Information, Reasoning, and AI 2024*,
Studies in Computational Intelligence 1280,
https://doi.org/10.1007/978-3-032-21143-9_7

We here entertain a related but different idea, to be referred to as a claim *Meaning Determines Valence (MDV)*:

(1) The meanings carried by the verbs of a language determine (within the general limits of the grammar) the verb valence system of the language.

MDV can be approached through at least two avenues. One is *The assumption of valence equivalence under translation equivalence* (AVEUTE):

(2) AVEUTE:

For any pair of languages A and B, modulo differences in the grammatical build-up of valence frames, if a verb VA in Language A is translation equivalent to a verb VB in Language B, then the valence frames of VB are the same as those of VA.

Another avenue is through exploring the question (3):

(3) Correlation Profiles:

In any given language, are there particular *aspects of verb meanings* which are reflected in particular *configurations of verb valence*, and vice versa?

Both avenues of investigation require large inventories of verb valences, and comprehensive systems of verb meaning specifications. These desiderata are met by a valence resource of Norwegian called *NorVal.*[1] Particular to its organization, a formal distinction is made between the notion of *all* the valence frames that a given verb can take, and a *single frame* as taken by a given verb. The notion of a single frame as taken by a verb is called a *lexically instantiated valence frame*, for short, a ***lexval***, while the notion of *all* the valence frames that a given verb can take, i.e., all its lexvals, is referred to as the ***valpod*** of the verb. The number of lexval tokens in NorVal is about 17,500, while, there being one valpod for each verb, the number of valpod tokens is 7500.

A subset of NorVal, to be referred to as ***NorVal-1952***, consists of 1952 verbs which carry 9722 of the lexvals of NorVal, and for these verbs, a system of semantics is encoded in terms of a *meta-level of 20 features* which encompasses all the verbs, in the sense of distinguishing nearly any verb from all the other verbs. These features are called **metafeatures**, abbreviated *MF*, or simply *features* when the meaning is clear. Just as a *valpod* defines the valence of a verb, the set of MFs defining the meaning of a verb is called its **sempod.**[2]

The verbs in NorVal-1952 being provided with English equivalents, the construction of an English valence-resource on the same format as *NorVal-1952*, tentatively called *EngNoVal*, was created through a procedure of 'beaming over' the valence inventory of NorVal-1952 and the MF specifications to the English counterpart by 'flipping' the Norwegian entry head ***VN*** of each NorVal-1952 entry with the English

[1] Cf. [10, 11].

[2] The full *NorVal* and *NorVal-1952* (as a text file '*Sempods-and-Valpods*') are located at: https://github.com/Regdili-NTNU/NorSource/tree/master/NorVal_files.

translation ***VE*** to serve as the head of an English entry, while keeping the MF specifications as its MF semantics, and the valpod of the Norwegian entry (modulo modifications due to grammar-differences between the languages) as the valpod of the English counterpart. AVEUTE (cf. (2)) is largely confirmed through this construction by the circumstance that about 70% of the resulting valpods are correct valpods for the English verbs.[3]

The procedure described does not explicitly address which semantic features correspond to which valence frames, since perceived synonymy between the verbs is the 'beaming' criterion, not specific identified properties of the individual verbs. What we entertain in the present investigation is the opposite, namely trying to identify which aspects of meaning give rise to which valence frames or sets of valence frames. This is thus the avenue (3), of finding Correlation profiles. The semantic metafeatures (MFs) will serve as the essential semantic system. The exploration here described concerns Norwegian, while a corresponding investigation for English could possibly be done along the same lines if based on EngNoVal.[4]

Section 2 presents the semantic metafeatures system, and further definitions of key notions of valence representation. Section 3 presents a first example of how meaning-valence relationships will presently be represented. Section 4 discusses the notion and representation of *subsenses* relative to the present system. Section 5 provides more extensive illustrations of meaning-valence relationships. Section 6 summarizes main points and gives a short outlook for possible further extensions.

2 Metafeatures, Sempods, Lexvals, Valpods, and Semvalpods

2.1 Semantic Features and Sempods

The linguistic system *Metafeatures (MF)* consists of about 20 features of verb meanings, designed with the aim that.

(i) all verbs in a language receive an appropriate characterization regarding content and commonalities/contrasts to other verbs, and
(ii) every verb in the language at the same time receives a characterization distinct from the characterization of any other verb.

The latter requirement might be trivially fulfilled by naming verb meanings simply by the name of the verb, however then failing to represent commonalities and contrasts regarding other verbs. As for requirement (i), analytic systems assigning four or five 'roles' such as 'agent', 'patient', 'theme' and 'experiencer' may fulfill the classificatory requirement, but such systems fall short of distinguishing each verb of a language from all the other verbs.

---

[3] Cf. [12].

[4] *EngNoVal* is also at https://github.com/Regdili-NTNU/NorSource/tree/master/NorVal_files.

The design of the features in MF seeks a compromise between classificatory expressivity of each feature and generality in coverage of each feature. The features of MF, first presented in [13], are as follows.[5]

(4) The MF Features of verb meaning:

'`Impact`': the verb activity leaves an impact in the domain of the activity.

'`Procedure`': the verb activity is characterized by a certain procedural pattern.

'`Directionality`': the verb activity unfolds directionally.

'`Perceptible`': the verb activity can be readily perceived.

'`ImpactSensory`': the verb expresses sensory effects impacting on a participant.

'`Perceiving`': the participant actively undertakes being in a position where she can receive sensory impact.

'`MedByPrcept`'(read as 'mediated by perception'): the act or event described is in some respect supported by perception or something perceived, but with a different focus than that of the above two categories.

'`Cogn`': the verb activity requires cognitive attention.

'`Dependence`': the verb activity presupposes or engenders structures of dependency.

'`Emotion`': the verb activity presupposes or engenders aspects of emotion.

'`Creation`': the verb activity has a creative contribution.

'`Response`': the verb activity typically serves in response to a previous event or established state.

'`Unfold`': the verb activity unfolds in time and is not punctual.

'`Perform`': the verb activity is in some respect willingly executed. As a particular type of performance, verbal performance is marked by the feature '`VerbalAct`'.

'`Domin`': the verb activity presupposes or engenders aspects of domination.

'`Poss`': the verb activity presupposes or engenders aspects of possession.

'`Criterial`': the verb carries *characteristic* features such as postures, patterns, event sequences, etc. They may be perceptual, thus representing a partial overlap with '`Perceptible`'.

'`SymbVal`': the verb activity includes aspects of symbolic functions, as could be expressed by sentential functions.

'`Loc`': the verb activity is specified for location.

'`Phys-Bod`': the verb activity unfolds in the physical or bodily domain.

'`PureRel`': the meaning of the verb does not pertain to any specific state of affairs, but rather represents general relations. (Using another term, the verb is not *situated*.)

The distribution of these features over the 1952 verbs is shown in Table 1.

[5] The list counts 22 features, but 2 or 3 of them have a more provisional status, for reasons stated at the end of 2.1. [13] focuses on whether the *size* of valpods can in any way be correlated with features of meaning of verbs. Here also some further features are defined: 'Sustain', 'Further', 'Imped', 'Prevent' and 'Init'. Being more oriented toward discourse structure, they are not considered presently.

Table 1 Number of verbs specified by each feature

Feature	Number of verbs
Perform	1401
SymbVal	1236
Procedure	1184
Impact	953
PhysBod	945
Cogn	937
Response	687
Domin	405
Criterial	391
Unfold'	390
VerbalAct	335
Creation	272
Emotion	226
Perceptible	204
Dependence	204
Directionality	185
MedByPrcpt	80
Poss	63
Loc	33
PureRel	27
Perceiving	26
ImpactSensory	14

The notation in which such features are used when representing a verb meaning is illustrated in (5). We call such a representation a ***sempod***.[6]

(5) {Directionality & Impact & Perform & PhysBod}

The MF-features in (4) are deployed in 1238 combinations, which is to say that there are 1238 different *types* of sempods instantiated across the 1952 verbs, while there are 1952 sempod *tokens* (with a proviso for the treatment of subsenses, cf. Sect. 4). Relative to the desideratum (ii) that "every verb in the language receives a characterization distinct from that of any other verb", for the 1952 verbs of NorVal-1952, 1238 verbs were uniquely distinguished in terms of the MFs, and relative to the 1300 verbs of EngNoVal, about 900 verbs were uniquely distinguished, thus about 65% in both cases. Relative to point (i), the coverage is 100%, in that all verbs get a specification, inherently reasonable as well as with a classificatory function. These may count as satisfactory scores relative to these desiderata, which suggests that

[6] What the expression (5) represents is formally a *set*, with '&' rather than commas used for enumeration.

the compromise sought in the use of MF features between specificity and generality fulfills its goal.

The features being at a higher level of generality than one would find in most systems of verb meanings, it is for instance not immediately obvious which verbs may belong under the sempod (5), even though, feature by feature, each feature could be seen as true of, e.g., the verb *shoot*. To be formally able to introduce a specification reflecting, e.g., the use of a *gun*, by which the notion of 'shoot' would presumably be better reflected, we outline below a way of embedding the 20 MF features in a hierarchical model of types and unification, where the MF features are construed as types and the type representing the unification of, e.g., the types represented in (5) would have subtypes defined down to a level of detail where an appropriate specification of *teeth* could be made. In essence, the MFs can be construed as *types* within a *Typed Feature Structures* (TFS) architecture, as we now outline.[7]

As a type, a metafeature can be associated with one or more *parameters*, that is, criteria by which one may judge whether the feature holds of a certain type of event or situation. In a TFS setting, such parameters can be represented as *attributes* declared by the types. For instance, two of the MFs in (5), construed as types, can be defined as having the attributes introduced in the type definitions in (6), the attributes stated in uppercase. Each attribute is declared with a *value*, namely a type, stated in italic lowercase, which is a top type in another type hierarchy, through which more fine-grained specifications can enter the specifications. (In (6), '*sit-type*' is the type dominating all the MF types; in the *tdl* encoding (cf. [15]), the symbol ': = ' means 'is a subtype of'.)

(6) *Directionality*: = *sit-type* & [DIRECTIONALITY *directed*].

Perform: = *sit-type* & [PERFORM-MODE *performance*, PERFORM-CRITERIAL-ITEM *thing-parameter*].

Directed, *performance* and *thing-parameter* being among the types introduced as values of the attributes mentioned in (6), these types will in turn each be heading a hierarchy of types, one sample type from each of these assumed hierarchies being stated in the right column in (7) (*dynMinTract* stands for 'dynamic, minimal trajectory'):

(7) Sample subtypes of

directed:... dynMinTract ...;

performance: ... willed ...;

thing-parameter: ... gun ...

The TFS system sustains unification, so that types like those corresponding to the MF features in (5) can be *unified*. The expression for the type resulting from that

[7] They can thereby be linked to the *LKB* platform (cf. [14, 15]) and be integrated in an LKB-based *grammar* within the framework HPSG (cf. [16])), such as the Norwegian computational grammar *NorSource* (cf. [17] and. [18]).

unification is written as (8), also in this hyphenated form to be referred to as a 'sempod':

(8) *Directionality-Impact-Perform-Physbod*

In this unification, the attributes introduced for the various MF types in (6) are accumulated, with the attribute-value pairs in (9) as parts of the structure; the attributes have as values the types in the left column in (7):

(9) [*Directionality-Impact-Perform-Physbod*

DIRECTIONALITY *directed,*

PERFORM-MODE performance,

PERFORM-CRIT-ITEM *thing-parameter,*

A subtype of this type closer to 'kick', provisionally called *Directionality-Impact-Perform-Physbod_1*, is defined by the AVM in (10), with the same attributes as in (9) but with values being the subtypes mentioned in the *right* column of (7).

(10) AVM for subtype of *Directionality-Impact-Perform-Physbod* suited for 'shoot':

[DIRECTIONALITY *dynMinTract*,

PERFORM-MODE *willed*,

PERFORM-CRIT-ITEM *gun*[8]].

The system of types and subtypes thereby assumed is summarized in (11):

(11) Summary of types and subtypes:

[8] Guns are of course not the only kind of thing with which one can shoot, thus here one will need a type subsuming various such items.

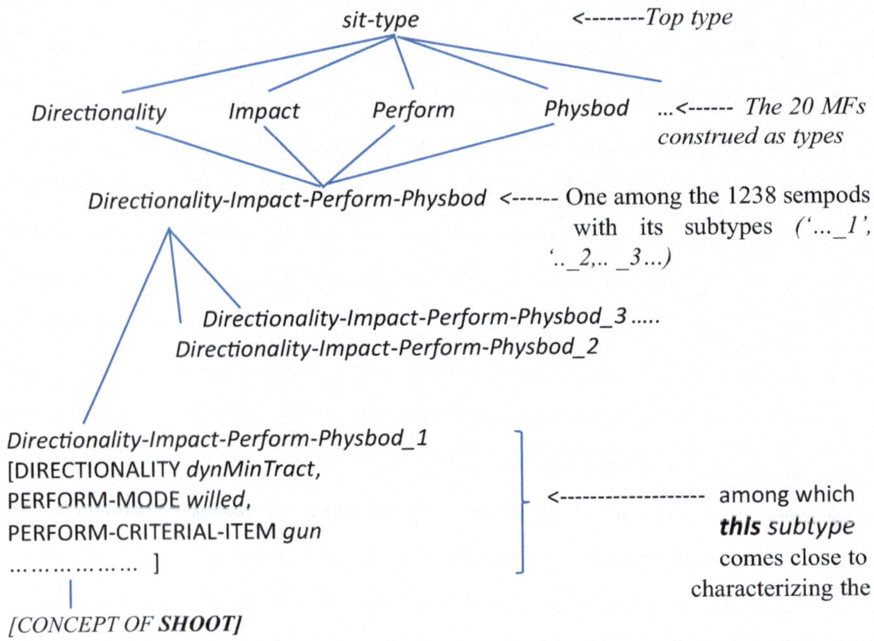

In the following, sempods will mainly be seen as *sets* of MF-features and less as unification objects. It may be noted, though, that the MF feature '`ImpactSensory`', defined as 'expressing sensory effects impacting on a participant', is rather to be seen as a sub-specification of the feature '`Impact`', 'lifted up' to MF-level for the present discussion. The same applies to the MF feature '`VerbalAct`', which can be construed as a subtype of '`Perform`', and partly also to the MF feature '`Criterial`', which states that there are criterial aspects to the notion in question. This is mainly a measure of convenience, since these aspects may differ from activity to activity, as anticipated in an attribute-name like 'PERFORM-CRITERIAL-ITEM' in (9 and 10), specific to criteria as concern *Performance*.

2.2 Lexvals and Valpods

The system of lexvals and valpods is laid out in much detail in [10, 11]).

A verb may in general have many valence environments. What we call a *lexval* is a valence frame as instantiated by a verb, thus a *lexically instantiated valence frame*, abbreviated ***lexval***. To illustrate the notion, a list of Norwegian lexvals is shown in (20), where what precedes the '&' in each example is the *lexval identifier*, and what succeeds the '&' is a short exemplifying sentence.

(12)

a. `sparke-på__intrObl-oblN` & han sparker på dørkarmen ('he kicks at the doorframe')
b. `sparke-ut__intrPrtcl` & keeperen sparker ut ('the goal keeper kicks out')
c. `sparke__intr` & babyen sparker ('the baby kicks')
d. `sparke__tr` & han sparker fangevokteren ('he kicks the prison guard')
e. `sparke__tr-obDir` & han sparker ballen vekk ('he kicks the ball away')
f. `sparke__trObl-oblPRTOFob` & hun sparker ham i benet ('she kicks him in the leg')
g. `sparke__trScpr-obRefl-scObNrgCsd-scPred` & han sparker seg svett ('he kicks himself sweaty')
h. `sparke__trScpr-scObCsd` & de sparker bordet i stykker ('they kick the table to pieces')

The lexval identifier consists of a *head*, which is a verb lexeme, and a *frame specification*, defined in terms of the CL annotation system. In (12), the lexeme *sparke* is in each case the head, while there are different frame specifications. In these frame specifications, `intrObl` in (12a) means 'intransitive plus an oblique PP' (and `trObl` means 'transitive plus an oblique PP'). The hyphenated part "`-på`" in (a) means that the preposition *på* is selected by the verb in this configuration; a similar convention is used for choice of particles, as for the particle *ut* 'out' in (12b). In a frame type specification, we distinguish between one initial '*global*' label, which indicates the overall type of the construction, and one or more *constituent* labels which define properties of constituents in the frame. The part `intrObl` in (12a) is a global label, and so is `trObl` in (f), while the part `oblPRTOFob` in the frame specification in (12f) is a constituent label, stating that the denotation of the oblique PP ('the leg') is part of the denotation of the object ('him'). The constituent label `trScpr` in the frame specifications in (g,h) means 'transitive with a secondary predicate', and the last constituent label `scObNrgCsd` in (g) gives a closer description relative to the part part `trScpr`, viz. that the secondary predicate is predicated of the object but not semantically an argument of the verb, and that the state of affairs denoted by the secondary predicate is caused by the action of the matrix verb. The second constituent label in (g), viz. `obRefl`, means that the direct object is a light reflexive.

As exemplified in (12a,b), attached to the lexval head there can be a hyphenated indication of a preposition or particle; in (12a), in `sparke-på`, `-på` represents the 'selected' preposition by which we say that this is an oblique constituent. The affixed `-på` representing categorial information, the string `-på__intrObl-oblN` thus is part of the *frame specification* of the lexval identifier `sparke-på__intrObl-oblN`. The part `intrObl-oblN` counts as the *frame type* involved, while we count the expression '`-på__intrObl-oblN`' as a *lexval* type, since the use of *på* as oblique-marker is an essential feature of the valence type inventory. Roughly assuming the number of such prepositions to be around 20 in Norwegian, there are thus *20 lexval types* corresponding to the *one frame type* `intrObl-oblN`.

A ***valpod*** is a condensed summary of all the lexval types that a verb may take, in the case of *sparke* being formally represented as in (13), illustrated in (12) above with examples[9]:

(13) {sparke:V__intr & V-på__intrObl-oblN & V-ut__intrPrtcl & V__tr-obDir & V__trObl-oblPRTOFob & V__trScpr-obRefl-scObNrgCsd-scPred & V__trScpr-scObCsd}

A ***valpod type*** resides in the set of *lexval types* constituting a valpod, *abstracting away the verb*; thus, the valpod type instantiated in (13) is (14):

(14) V__intr & V-på__intrObl-oblN & V-ut__intrPrtcl & V__tr-obDir & V__trObl-oblPRTOFob & V__trScpr-obRefl-scObNrgCsd-scPred & V__trScpr-scObCsd

Among the 7500 valpods in NorVal, the number of valpod *types* is 1914, counting valpods in terms of lexval types, thus with specifications of selected prepositions and particles included.

2.3 Semvalpods

A ***semvalpod*** is a combination of a sempod with a valpod. Illustrating with the Norwegian verb *sparke* 'kick', its semvalpod is written in text format as in (15a) and in tabular format as in (15b).

(15) Semvalpod of *sparke* 'kick'

a. Text format

sparke ('kick'): `{Impact & Perform & Criterial & Loc & PhysBod}` & VALPOD: `{sparke:V__intr & V-på__intrObl-oblN & V-ut__intrPrtcl & V__tr-obDir & V__trObl-oblPRTOFob & V__trScpr-obRefl-scObNrgCsd-scPred & V__trScpr-scObCsd}`_7-items.[10]

[9] 'V-på__intrObl-oblN' and '-på__intrObl-oblN' both count as *lexval types*, in terms of the terminology laid out above.

[10] In most valpod files, the valpod specifications are consistently ended with the number of lexvals contained, as here: '7 items'.

b. Tabular format

Verb entry	Sempod	Valpod	Examples (aligned with the lexvals in the column to the left; cf. (14) for glosses)
sparke ('kick')	{Impact & Perform & Criterial & Loc & PhysBod}	{sparke: V-på__intrObl-oblN & V-ut__intrPrtcl & V__intr & V__tr-obDir & V__trObl-oblPRTOFob & V__ trScpr-obRefl-scObNrgCsd-scPred & V__trScpr-scObCsd}	han sparker på dørkarmen & keeperen sparker ut & babyen sparker & han sparker den vekk & hun sparker ham i låret & han sparker seg svett & de sparker bordet i stykker

NorVal-1952 consists of 1952 semvalpod tokens. In each, the *sempod* part is one among the 1238 types of sempods, and the *valpod* part is one among the 1914 types of *valpods*. From a formal viewpoint one may also conceive of a larger space of possible semvalpods, namely where the sempods are all logically possible combinations among the 20 MFs, and the valpods are all logically possible combinations among the 340 *frame types*, or even, all combinations among the possible *lexval types*. The latter possibility is summarized in the first line of Table 2, while the attested set of possibilities are those in the second line. We refer to the latter as *lexically anchored* semvalpods, and their constituents as *lexically anchored sempods* and *lexically anchored valpods*.

In investigating correlations between meaning and valence, since both sides are formalized in terms of sets, we have the formal possibility of exploring correlations in terms of *subsets* from either side. Thus, we allow for conceiving of MF combinations which are subsets of actual sempods but not necessarily by themselves constituting actual sempods, and likewise for lexval combinations. Such subsets may be called *schemata* of sempods and valpods, respectively; together they can constitute a *semvalpod schema*. A schema generally will be said to *match* or *unify with* any larger set of which it is a subset, and if such a larger set is lexically anchored, we say that the lexically anchored set *instantiates* the schema.

Table 2 Numbers concerning sempods, valpods and semvalpods

	Sempod types	Valpod types	Semvalpod types
Formally possible	The power set of the semantic features in (4)	The power set of the 340 frame types, restricted to subsets with less than 30 lexvals	Cartesian product of the power sets mentioned in the columns to the left
Attested as 'lexically anchored' in Norwegian	1238 types in NorVal-1952	1914 types in NorVal-1952	1952 types and tokens in NorVal-1952

3 Correlations Between Meaning and Valence

In the terms now established, patterns to be investigated can typically be covered by the following types of statement:

(16) Types of Correlation statements:
 (a) *Semantically Based Statement:*

For a sempod *S* and a semvalpod *SVP* containing *S*, if *S* contains features *a, b, c,* …then *SVP* will have the valpod *VP* or the sub-valpods *VP1* or *VP2* or *VP3*, or lexvals *L1, L2, L3* common to the valpods.

b *Valence Based Statement:*

For a valpod *VP* and a semvalpod *SVP* containing *VP*, if *VP* contains lexvals *L1, L2, L3,* … then *SVP* will have a sempod *S* with sub-sempods *S1, S2,*… or containing features *a, b, c.,* … common to the sempods.

We first illustrate (16a). As the sempod *S* we consider a subset of the sempod in (15), viz. the set in (17), to be referred to as *The Four Features*:

(17) *The Four Features*

`{Impact & Perform & Criterial & PhysBod}.`
There are 182 semvalpods containing a sempod with these features. To have a set of more manageable size, we distinguish 39 of these semvalpods by the arbitrary criterion that the four features occur contiguously to each other; we refer to this set as "*The 39 semvalpods"*; cf. Appendix 3.

The valpods in "*The 39 semvalpods"* are mostly between 5 and 10 lexvals in size, and at first sight rather heterogeneous; we refer to these valpods as "*The Valpods-of-the-39"*. Rather than try exploring *The Valpods-of-the-39* directly, we first try to find lexvals *common* to them, i.e., the last option in (16a). This is shown in Table 3. For each lexval type entered in the leftmost column, column 2 states the number of instances of that type in "*The Valpods-of-the-39"*. For comparison, columns 3 and 4 state the number of instances of the type in the NorVal-1952, in column 3 in absolute numbers and in column 4 in proportion to the total number of the verbs in NorVal-1952. Column 5 compares the instances in column 2 proportion-wise to the numbers in column 4, giving an indication of which lexvals are *typical* of "The Valpods-of-the-39"—the higher the number in column 5, the higher the typicality. (Note that what is here entered as 'PRTOF' in most cases instantiates as 'V__trObl-oblPRTOFob', while there are also occurrences of '... oblPRTOFsu'.)

Standing out as particularly 'typical' is the occurrence rate of frames with 'PRTOF', here being 15 times more frequent than its average in NorVal-1952.

Shifting to the first option in (16a), i.e., identifying a set of lexvals (*L1, L2, L3*) constituting a common subset of a set of valpods, we use the three most 'typical' lexvals according to Table 3 as forming a possible common subset, which we call "*The Three Lexvals*":

(18) "*The Three Lexvals*":

V__trPrtcl & V__trScpr & ...PRTOF

"*The Four Features*" and "*The Three Lexvals*" are *schemata* of sempods and valpods, respectively, in the terms suggested above. The semvalpod schema constituted by the combination of "The Four Features" and "The Three Lexvals" is instantiated by only one lexically anchored semvalpod, namely for the verb *skyte* 'shoot', as opposed to the 39 lexically anchored semvalpods instantiating "*The Four Features*" alone. If we remove PRTOF from "*The Three Lexvals*", semvalpods instantiating 4 verbs result—for *bite* 'bite', *nappe* 'pinch', *skyte* 'shoot', and *stille* 'posit', and if instead we remove V__trPrtcl from "*The Three Lexvals*", the resulting semvalpods instantiate the semvalpods for *skyte*, *slikke* 'lick' and *sparke*; the figures are summarized in Table 4.

This illustrates the kinds of findings we may get under the course (16a) of queries. If we start from the valence side, i.e., (16b), and see how many semvalpods match *The Three Lexvals* (cf. (18)), if no semantic parameters are supplied, the number of instantiated verbs is 6, viz., for the verbs *komme* 'come' (which by itself has 30 lexvals), *skjære* 'cut' (which has 12 lexvals), *skyte* 'shoot' (with 12 lexvals), *slå* 'beat, hit' (with 25 lexvals), *ta* 'take' (with 44 lexvals), and *tørke* 'dry' (with 9 lexvals), thus verbs with in part extremely large valpods. If we reduce the search further down to the pair 'V__trPrtcl & V__trScpr' the number of matching semvalpods as expected is larger, viz. 55, and for cross-classifying with semantic features we could go through the resulting semvalpod sets analogously to above, giving 44 semvalpods when combining with Perform, 16 semvalpods when further combining with Criterial, etc., down to semvalpod schema 3 of Table 4 when all of *The Four Features* are included. Thus, searches over 'pods' in the meaning and the valence dimension here give an approximation to a 'butterfly'-like shape, with 1238 lexically anchored sempods as the left wing and 1914 lexically anchored valpods as the right wing, and with the 'body' consisting of a semvalpod schema, thus with a sempod and a valpod not necessarily instantiated as the sempod or valpod of any lexically anchored semvalpod.

We explore in Sects. 4.2 and 5 examples of what can putatively be called 'semantic types of verbs', where expectations of possible correlations may be tested. Clusters

Table 3 'Typicality' profile for the lexvals of *"The Valpods-of-the-39"*

1	2	3	4	5
Lexval type	Instances of the lexval in *"The Valpods-of-the-39"*	*Reference values*: Instances of the lexval in the 1952 selection from NorVal	*Reference values*: Dividing *Reference values* in column 3 by 50, to get appx number proportional to those for *"The Valpods-of-the-39"*	Proportion of instances of lexval in *"The Valpods-of-the-39"* compared to number of verbs in the 1952 selection from NorVal: '3' means '3 times as much', i.e., 3 times as much as average in NorVal-1952
V__trPrtcl	50	813	16	3.1
V__ trObl-oblN	33	564	12	2.6
V__ intrObl-oblN	30	987	18	1.8
V__tr	22	1313	26	1
V__intr	20	737	17	1
… PRTOF …	20	87	1.6	15
V__intrPrtcl	16	363	7	2
V__tr-obDir	13	107	5	2.8
V__trScpr	11	157	3	3.9
V__tr-obRefl	10	354	7	1.3
V__ trPrtcl-obRefl	7	129	6	1
V__intr-suDir	6	123	6	1
… oblDECL	5	515	10	0.5
… oblINTERR	3	365	7	0.5
…Inf	0	1072	20	0

(continued)

Table 3 (continued)

1	2	3	4	5
…Bareinf	0	21	0.4	0
…obDECL	0	405	20	0
-obINTERR	0	246	5	0

Table 4 Some semvalpod schemata

	Sempod	Valpod	Number of verb instances
semvalpod schema 1	Impact & Perform & Criterial & PhysBod	V__trPrtcl & V__trScpr & …PRTOF	*1: skyte*
semvalpod schema 2	Impact & Perform & Criterial & PhysBod	V__trScpr & …PRTOF	*3: skyte, slikke, sparke*
semvalpod schema 3	Impact & Perform & Criterial & PhysBod	V__trScpr & V__trPrtcl	*4: bite, nappe, skyte, stille*

of similar 'butterflies' in the above terms may show that some semantic features are more constitutive than others in determining valpod structures, and conversely, thus what is envisaged in the exploration avenue (3) in the Introduction.

We first recognize an essential limitation to the term '*the* meaning of a verb', namely that a verb often has many shades of meaning, also called 'contextually determined variants', for which we here use the notion *subsense*. The next section outlines a format for the analysis of subsenses, compatible with the current format.

4 Subsenses

4.1 Representing Subsenses

For the verb *fryse* 'freeze, be/feel cold', three subsenses can be recognized:

(19) Illustrations of subsenses for *fryse*:

`fryse_SUB_EXP` & ('be cold'), as in *tilskuerne fryser* 'the spectators feel cold';
`fryse_SUB_AG` & ('freeze'), as in *de fryser filetene* 'they freeze down the fillets';
`fryse_SUB_UND` & ('freeze'), as in *vannet fryser* 'the water freezes'.

The suffix '`SUB`' here means being a subsense of the verb in question, and the further suffixes '`EXP`', '`AG`' and '`UND`' intuitively keep the subsenses apart. Formalizing these subsenses, they are given the following specifications, using the *sempod* format of semantic representation:

(20) Sempods for subsenses of *fryse*:

`fryse_SUB_EXP` & ('feel cold') & `{ImpactSensory & Response & Unfold & PhysBod}` .
`fryse_SUB_AG` & ('freeze') & `{Perform & Unfold & PhysBod}`.
`fryse_SUB_UND` & ('freeze') & `{Response & Unfold & PhysBod}`.

To some extent, also syntactic patterns reflect subsenses, as illustrated by the following sets of examples for each of the above subsenses, encoded in terms of valpods:

(21) Valpods specific to subsenses of *fryse*:

fryse_SUB_EXP:

Examples: *tilskuerne fryser* 'the spectators feel cold' & *tilskuerne fryser på bena* 'the spectators are cold on their feet'.

Valpod patterns: fryse: {V__intr & V__intrObl-oblPRTOFsu}.
fryse_SUB_AG:

Examples: *de fryser filetene* 'they freeze the fillets' & *vi fryser ned filetene* 'they freeze down the fillets' & *vi fryser ut protestantene* 'we freeze out the protesters'.

Valpod pattern: fryse: {V__tr & V-ned__trPrtcl & V-ut__trPrtcl}.
fryse_SUB_UND:

Examples: *vannet fryser* 'the water freezes' & *innsjøen fryser til* 'the lake freezes in'.

Valpod pattern: fryse: {V__intr & V-til__intrPrtcl}.
As the third example of fryse_SUB_AG indicates, the distinction between 'literal' and 'metaphorical' is not reflected in this tentative subsense inventory.

It should be noted that V__intr here obtains in two subsense specifications, and likewise V__tr; thus, a given lexval can be shared across different subsenses. In other words, differentiation between lexvals does not systematically reflect differentiation among subsenses.

Combining the aspects of specification shown in (20) and (21) into semvalpods, the semvalpods of *fryse* will look as follows for its three subsenses:

(22) Subsense-aligned semvalpods for *fryse*:

a. fryse_SUB_EXP & ('feel cold') & {ImpactSensory & Response & Unfold & PhysBod} & VALPOD: fryse: {V__intr & V__intrObl-oblPRTOFsu} *tilskuerne fryser* & *tilskuerne fryser på bena* & ENDex[11]
b. fryse_SUB_AG & ('freeze') & {Perform & Unfold & PhysBod} & VALPOD: fryse: {V__tr & V-ned__trPrtcl & V-ut__trPrtcl} & EX: de fryser filetene & *vi fryser ned filetene* & *vi fryser ut protestantene* & ENDex
c. fryse_SUB_UND & ('freeze') & {Response & Unfold & PhysBod} & VALPOD: fryse: {V__intr & V-til__intrPrtcl} *vannet fryser* & *innsjøen fryser til* & ENDex

[11] For NorVal-1952, cf. [13]), and for NorVal, cf. [10, 11].

4.2 An Illustration

Subsense differentiations apply to nearly all the 14 verbs carrying the feature ImpactSensory,[12] exemplified in Table 5. Most of the semvalpods here represent subsenses of the verbs in question (thus, first on the list is the subsense (22a) of *fryse),* hence most of the entries are accompanied by a' _SUB'-affixed alternant, even if it is the only subsense of the verb included in this list. Note that homonyms are distinguished by a hyphenated number, thus the '−2' in 'kile-2' serves to distinguish this verb from another with the same spelling. Each entry is followed by examples illustrating the lexvals in the valpod, in the same order.

We comment in Sect. 5 on various aspects of these semvalpods in comparison with other semvalpods to be described.

4.3 'Main Sense' and 'Subsense'

Having a format for representing subsenses, the introduction of subsenses in NorVal-1952 is still at its beginning. In the original NorVal-1952, where subsenses are not marked at all, the representation of *fryse* is as follows:

(23) Semvalpod representation of *fryse* in NorVal-1952:

fryse & ('freeze') & {ImpactSensory & Response & Perform & Unfold & PhysBod} & VALPOD: {fryse:V__intr & V__intrObl-oblPRTOFsu & V__tr & V-ned__trPrtcl & V-til__intrPrtcl & V-ut__trPrtcl}.

Comparing (23) with the subsense representations in (22), repeated, both the sempod and the valpod parts of (23) can be seen as being the set-*unions* of the respective sempod and valpod representations in (22).

(22) (repeated)Subsense-aligned semvalpods for *fryse*:

a. fryse_SUB_EXP & ('feel cold') & {ImpactSensory & Response & Unfold & PhysBod} & VALPOD: fryse: {V__intr & V__intrObl-oblPRTOFsu} *tilskuerne fryser & tilskuerne fryser på bena* & ENDex
b. fryse_SUB_AG & ('freeze') & {Perform & Unfold & PhysBod} & VALPOD: fryse: {V__tr & V-ned__trPrtcl & V-ut__trPrtcl} & EX: de fryser filetene & *vi fryser ned filetene & vi fryser ut protestantene* & ENDex
c. fryse_SUB_UND & ('freeze') & {Response & Unfold & PhysBod} & VALPOD: fryse: {V__intr & V-til__intrPrtcl} *vannet fryser & innsjøen fryser til* & ENDex

[12] As noted at the end of Sect. 2.1, the content of ImpactSensory might be better construed as a subtyoe of the feature Impact. For the considerations to follow, however, this is immaterial.

Table 5 Sempods and valpods for verbs with 'ImpactSensory' as semantic feature

	Sempods	Valpods	Examples
fryse_SUB_EXP ('be cold')	{ImpactSensory & Response & Unfold & PhysBod}	{fryse:V__intr & V__intrObl-oblPRTOFsu}_2-items	tilskuerne fryser & tilskuerne fryser på bena
kile-2_SUB_EXP ('tickle')	{ImpactSensory & Impact & Criterial & PhysBod}	{kile-2:V__impersObl-oblPRTOFexper & V__tr & V__trObl-oblPRTOFob}_3-items	det kiler i nesen & tråden kiler henne i ansiktet & stoffet kiler huden
kile-2_SUB_AG ('tickle')	{ImpactSensory & Impact & Perform & Criterial & PhysBod}	{kile-2:& V__tr & V__trObl-oblPRTOFob & V__trObl-obRefl-oblPRTOFob}_3-items	gutten kiler henne & gutten kiler henne under foten & hun kiler seg under nesen
kjenne_SUB_EXP ('feel')	{ImpactSensory & MedByPrcept & Cogn & Impact & Response & SymbVal & PhysBod}	{kjenne:V__trScpr-obRefl-scObNrg-scBareinf & V__trScpr-scObNrg-scBareinf}_2-items	hun kjenner seg stivne & jeg kjenner det stivne
kjøle_SUB_EXP ('cool')	{ImpactSensory & Impact & Unfold & PhysBod}	{kjøle:V__tr}_1-item	vinden kjøler pannen
kjøle_SUB_AG ('cool')	{ImpactSensory & Impact & Perform & Unfold & PhysBod}	{kjøle:V__tr & V__tr-obRefl & V-ned__trPrtcl & V-ned__trPrtcl-obRefl}_4-items	politiet kjøler gemyttene & hun kjøler seg & vi kjøler ned fisken & hun kjøler seg ned
lukte_SUB_AG ('smell')	{Perceiving & ImpactSensory & Response & Perform & Criterial & SymbVal & PhysBod}	{lukte:V-på__intrObl-oblN}_1-items	katten lukter på grøten
lukte_SUB_COG ('smell')	{Perceiving & ImpactSensory & Cogn & Response & Perform & Criterial & SymbVal & PhysBod}	{lukte:V__tr & V__tr-obDECL & V__tr-obINTERR}_3-items	hun lukter blomstene & hun lukter at noen har vært der & hunden lukter om det har vært katter der

(continued)

Table 5 (continued)

	Sempods	Valpods	Examples
lukte_SUB_ EXP ('smell')	{Perceiving & ImpactSensory & Cogn & Response & Criterial & SymbVal & PhysBod}	{lukte:V__impers & V__ impersAdv & V-av__ impersObl-oblN & V__intrAdv & V__tr & V__trImpers & V-av__ trImpersObl-oblN}_7-items	det lukter & det lukter godt & buketten lukter krysantemum & det lukter av barnåler & blomstene lukter godt & det lukter barnåler & det lukter dårlig samvittighet av ham
smake_SUB_ AG ('taste')	{Perceiving & ImpactSensory & Response & Perform & Criterial & SymbVal & PhysBod}	{smake:V-på__intrObl-oblN}_ 1-items	
smake_SUB_ COG ('taste')	{Perceiving & ImpactSensory & Cogn & Response & Perform & Criterial & SymbVal & PhysBod}	{smake:V__tr}_1-items	hun smaker tilsetningene
smake_SUB_ EXP ('taste')	{Perceiving & ImpactSensory & Cogn & Response & Criterial & SymbVal & PhysBod}	{smake:V__impers & V__ impersAdv & V-av__ impersObl-oblN & V__intrAdv & V__tr & V__trImpers & V-av__ trImpersObl-oblN}_7-items	det smaker & det smaker godt & det smaker av sardiner & suppen smaker godt & suppen smaker buljong & det smaker tran & det smaker tran av suppen
smerte_SUB_ EXP_ SENSORY ('hurt')	{ImpactSensory}	{smerte:V__ impersObl-oblPRTOFexper}	det smerter i armen
stikke-1_ SUB_EXP ('sting')	{ImpactSensory & Perceptible & Impact & Criterial & PhysBod}	{stikke-1:V__ impersObl-oblPRTOFexper & V_ _intr & V__tr & V__ trObl-obRefl-oblPRTOFob & V-på__trObl-obRefl-oblN}_ 5-items	det stikker i albuen & tornene stikker & tornene stikker sauene & prinsessen stikker seg i armen & prinsessen stikker seg på nålen &

(continued)

Table 5 (continued)

	Sempods	Valpods	Examples
stikke-1_ SUB_AG ('sting')	{ImpactSensory & Perceptible & Impact & Perform & Criterial & PhysBod}	{stikke-1:V-med__intrObl-oblN & V__tr-obDir & V__ trObl-oblPRTOFob & V-med__ trObl-oblN}_4-items	
stinke_SUB1_ EXP ('stink')	{ImpactSensory & Unfold & Criterial & PhysBod}	{stinke:V-av__impersObl-oblN & V__intr & V-av__intrObl-oblN}_ 3-items	det stinker av fiskelim & fisken stinker & huset stinker av fisk
svartne ('blacken')	{ImpactSensory & Impact & Response & Criterial & PhysBod} &	{svartne:V__impers & V-for__ impersObl-oblN & V__intr}_ 3-items	det svartner & det svartner for øynene & det svartner for ham & horisonten svartner
svi_SUB_ EXP ('burn, suffer')	{ImpactSensory & Impact & Unfold & Criterial & PhysBod}	{svi:V__impers & V__ impersObl-oblPRTOFexper & V_ _intr} 3-items	det svir & det svir i hodebunnen & hodebunnen svir
valne ('become numb')	{ImpactSensory & Perceptible & Impact & Response & Domin & Criterial & Loc & PhysBod}	{valne:V__intr & V__tr & V-bort__intrPrtcl}_3-items	fingrene valner bort & fingrene valner & kulden valner fingrene
varme_SUB_ EXP ('heat')	{ImpactSensory & Perceptible & Impact & Procedure & Unfold & PhysBod}	{varme:V__ impersObl-oblPRTOFexper &}_ 1-items	det varmer på tærne
verke_SUB_ EXP ('ache')	{Further & ImpactSensory & Impact & Emotion & Response & Domin & Criterial & PhysBod}	{verke:V__ impersObl-oblPRTOFexper & V_ _intr}_2-items	det verker i leddene & tannen verker

The relation between (23) and (22) can thus be construed as a 'subset extraction' from (23), with the possibility of overlapping subsets. This being a logical relationship, we can also conceive it as a *procedural* relationship for systematically turning semvalpods into subsense-aligned semvalpods. A record of such a procedure for the present case is indicated in (24), with (24a) as the 'source' structure, and (b-d) as 'extracted' stages, from (a) to (b), from (a) to (c), and from (a) to (d):

(24) A 'subsense record' representation of the lexeme *fryse*, with the 'inclusive' semvalpod (a), and the 'extracted' semvalpods (b-d):

a. `fryse & ('freeze') & {ImpactSensory & Response & Perform & Unfold & PhysBod} & VALPOD: fryse: {V_ _intr & V__intrObl-oblPRTOFsu & V__tr & V-ned_ trPrtcl & V-til__intrPrtcl & V-ut__trPrtcl}`

b. fryse_SUB_EXP & ('feel cold') & {ImpactSensory & Response & Unfold & PhysBod} & VALPOD: fryse: {V__intr & V__intrObl-oblPRTOFsu} *tilskuerne fryser* & *tilskuerne fryser på bena* & ENDex
c. fryse_SUB_AG & ('freeze') & {Perform & Unfold & PhysBod} & VALPOD: fryse: {V__tr & V-ned__trPrtcl & V-ut__trPrtcl} & EX: *de fryser filetene* & *vi fryser ned filetene* & *vi fryser ut protestantene* & ENDex
d. fryse_SUB_UND & ('freeze') & {Response & Unfold & PhysBod} & VALPOD: fryse: {V__intr & V-til__intrPrtcl} *vannet fryser* & *innsjøen fryser til* & ENDex

For the purpose of finding correspondences between meanings and valences, the representations (24b-d) state correspondences exactly, so, although they represent added complexity in terms of number of lines, they add to the adequacy of the resource. In this respect, in contrast, (24a) would be just a misleading extra layer. Thus, for the project of exploring correlations between meaning and valence, one might want to exclude the 'inclusive' semvalpod.

From the more general perspective of building a valence resource, however, the 'inclusive' semvalpod may have a function as a summary of sempods and valpods relevant to a verb, and thus be used as a representation of the 'lexeme per se' in question, when subsenses are not in focus.

An alternative for representing the 'lexeme per se' could be the *intersection* of the subsense-aligned sets. An immediate counter-consideration to this would be that the 'intersection' in the case of (24b-d) would be simply {Unfold & PhysBod}, which is hardly a representation of 'freezing'/ 'being cold'. In principle there might also be cases where the subsense sempods of a verb are connected only through 'family resemblance' and with only an empty intersection.

Retaining both 'inclusive' and 'sense-specific' semvalpods, a question will still be if they could be better marked than in the illustration so far, for instance to be visible for automatic procedures for skipping the inclusive semvalpods for verbs where such a distinction has been marked. This issue remains to be addressed.

It should be noted that the present notion of a *subsense of a verb* is distinct from the notion of a *subtype of the type representing the meaning of a verb*, as this was construed in Sect. 2.1. Two subsenses of a verb are represented by different sempods, and in the construal of sempods in terms of types, different top types in the type hierarchies represent the different senses. Thus, as outlined in Sect. 2.1 (cf. the hierarchy in (11)), when the sempod *Directionality-Impact-Perform-Physbod_1* is a subtype of the sempod *Directionality-Impact-Perform-Physbod*, that will not mean that one is a subsense of the other, or that they represent subsenses of the same verb.

This ends our remarks on subsense representations, and we return to semantics-valence correlations.

5 Correspondences Meaning—Valence

Table 5 in the previous section may also be seen as an illustration of how a single semantic feature may serve as a starting point—also called a 'query feature'—for finding a verb profile, a possibility suggested at the end of Sect. 3. The opposite, i.e., querying with many features, is a query in terms of what we call a *sempod schema* or *lexval schema*, named throughout with expressions such as "The Four Features" or "The Three Lexvals". The present section follows both lines, with Sect. 5.1 drawing observations from the profiles of two 'low coverage' search features, while in Sect. 5.2 we consider profiles resulting from a feature with very high coverage. We will see that low coverage features can function reasonably well in single feature queries, while high and medium coverage features can be used as query features only, or preferably, as part of sempod schemata or lexval schemata.

5.1 *Low-Coverage Features as Single Query Features*

We compare profiles of '`ImpactSensory`', applied as a single query feature, with profiles of '`Perceiving`', also applied as a single query feature.

5.1.1 Profiles for *ImpactSensory*

Lexvals cooccurring in semvalpods with `ImpactSensory` in Table 5 are singled out individually in Table 6, also showing the number of instances in Table 5.

Semantic features cooccurring with `ImpactSensory` in the sempods in Table 5 are shown in Table 7:

Standing out among the lexval types in Table 6 is the lexval type '`V__impersObl-oblPRTOFexper`', where the part '`exper`' in the label string indicates a part-whole relationship relative to an *implicit* experiencer. This lexval type has no other appearances in the total inventory. (25) shows the sempods in its semvalpod environments.

(25) Sempods cooccurring with the lexval `V__impersObl-oblPRTOFexper`:

```
ImpactSensory & Criterial & PhysBod
ImpactSensory & Perceptible & Criterial & PhysBod
ImpactSensory & Perceptible & Procedure & Unfold & PhysBod
ImpactSensory & Emotion & Response & Domin & Criterial & PhysBod
```

When comparing with the overall numbers in NorVal-1952 as depicted in Table 7, rightmost column, one notices the absence of `Cogn`, `SymVal` and `Perform` in (25), while, on the other hand, there is a higher-than-average presence of `PhysBod` and `Criterial` in (25).

Table 6 Lexval instantiations relative to `ImpactSensory`

Lexval type	Instances in semvalpods in Table 5
V__tr	9
V__intr	7
V__impersObl-oblPRTOFexper	6
V__impersObl-oblN	4
V__intrObl-oblN	4
V__impers	3
V__trPrtcl	3
V__trObl-oblPRTOFob	3
V__trImpersObl-oblN	2
V__trObl-obRefl-oblPRTOFob	2
V__impersAdv	2
V__trImpers	2
V__intrAdv	2
V__intrPrtcl	1
V__tr-obDECL	1
V__tr-obINTERR	1
V__trScpr-obRefl-scObNrg-scBareinf	1
V__trScpr-cObNrg-scBareinf	1
V__tr-obRefl	1
V__trPrtcl-obRefl	1
V__intrObl-oblPRTOFsu	1

Table 7 Semantic features occurring in sempods together with `ImpactSensory` in Table 5

Feature	Instances of the feature's cooccurrence with `ImpactSensory` in Table 5	Instances of the feature in NorVal-1952 overall
PhysBod	21	945
Criterial	15	391
Response	12	687
Perform	7	1401
SymbVal	7	1236
Unfold	6	390
Cogn	5	937
Emotion	1	226

We next show three valpod environments for V__impersObl-oblPRTOFexper:

(26) Valpod environments of V__impersObl-oblPRTOFexper:

V__tr & V__trObl-oblPRTOFob
V__intr & V__tr & V__trObl-obRefl-oblPRTOFob & V-på__trObl-obRefl-oblN
V__intr

The two PRTOF-lexvals here indicate an explicit experiencer expressed as object; together the occurrences of explicit and implicit PRTOF-lexvals comply with an idea of bodyparts being essential when it comes to sensory impact.

Both the typicality profiles for semantic features and lexvals (Tables 6 and 7), and typicality profiles for cooccurring items ((25) and (26)), point to possible convergences corresponding to a conceptual notion of 'sensory impact'. With a proviso that the set of verbs corresponding to this feature is exceptionally small in terms of the distributions shown in Table 1 (viz., 14 verbs), the observations illustrate what it could mean that semantic features possibly are constitutive in determining valpod structures.

5.1.2 Profiles for *'Perceiving'*

The feature Perceiving applies to a set of verbs slightly larger than the set associated with ImpactSensory. The semvalpods specified for it as a query feature are shown in Appendix 1. We display profiles for Perceiving in the same way as done for ImpactSensory above, first for lexvals as shown in Table 8. For comparison with ImpactSensory, the figures from Table 6 are repeated. Lexvals particularly to be noted in comparative respects are rendered in boldface:

Table 9 shows which semantic features cooccur with Perceiving in the semvalpods, again comparing with ImpactSensory, now with figures repeated from Table 7; here also overall numbers for NorVal-1952 are shown, repeated from Table 1.

Perform, Cogn, SymbVal are features significantly more activated when cooccurring with Perceiving in a sempod than when cooccurring with ImpactSensory. The higher number of instances of Cogn and SymbVal can be correlated with the high number of lexvals with DECL and INTERR in Table 8. The higher number of IntrObl-oblN for Perceiving could go together with the higher number of V__tr for Perceiving, both compatible with the feature Perform. From the viewpoint of correlation between meaning and valence frames, it is thus possible that we can make a similar tentative summary here as in the previous case.

Table 8 Lexval type instantiations relative to `Perceiving`

Lexval type	Instances in semvalpods containing `Perceiving`	Instances in semvalpods in Table 6 containing `ImpactSensory`
V__tr	15	9
V__intr	8	7
V__impersObl-oblPRTOFexper	6	6
V__impersObl-oblN	0	4
V__intrObl-oblN	**22**	**4**
V__impers	2	3
V__trPrtcl	4	3
V__trObl-oblPRTOFob	0	3
V__trImpersObl-oblN	4	2
V__trObl-obRefl-oblPRTOFob	0	2
V__impersAdv	2	2
V__trImpers	2	2
V__intrAdv	2	2
V__intrPrtcl	1	1
V__tr-obDECL	**11**	**1**
V__tr-obINTERR	**12**	**1**
V__trScpr-obRefl-scObNrg-scBareinf	2	1
V__trScpr-cObNrg-scBareinf	2	1
V__tr-obRefl	1	1
V__trPrtcl-obRefl	0	1
V__intrObl-oblPRTOFsu	0	1
V__intrObl-oblINTERR	**5**	
V__intrObl-oblDECL	**7**	
V__intrObl-oblEqSuInf	1	

5.2 *High-Coverage Feature:* `SymbVal`

`SymbVal` is a feature indicating that the act or state in question has a symbolic element as part of its unfolding, so that committing the act or state entails using a *symbol*. A typical instance is an act of *utterance*, where the utterance of a *sentence* is an integral part.

Most of the features in the present system are instantiated only within large sets of verbs, too large to allow one to easily get an impression about commonalities and differences just by inspection. `SymbVal` is such a feature. It appears in 1236 semvalpods—let's call them *"The SymbVal Semvalpods"*. For comparison, while the full number of lexvals in NorVal-1952 is *9726*, the full number of lexvals in "*The*

Table 9 Semantic features cooccurring with `Perceiving`

Feature	Instances of the feature's cooccurrence in a sempod with `Perceiving`	Instances of the feature's cooccurrence in a sempod with `ImpactSensory`	Instances of the feature in NorVal 1952 overall
PhysBod	27	21	945
Criterial	16	15	391
Response	10	12	687
Perform	**21**	**7**	1401
SymbVal	**14**	**7**	1236
Unfold	7	6	390
Cogn	**16**	**5**	937
Emotion	2	1	226
Impact	4		
ImpactSensory	6	23	

SymbVal Semvalpods" is *6257*. A particular feature of these symvalpods is the use of *clausal lexvals.* A lexval counts as *clausal* when any of the following partial label strings occurs in its specification:

`DECL` (for declarative subordinate clause),
`Eq` (for 'Equi'-controlled infinitival clause),
`INTERR` (for interrogative subordinate clause),
`Absinf` (for 'absolute'-interpreted infinitival clause),
`Bareinf` (for infinitival clause without infinitive marker).

Table 10 shows the most frequent lexvals containing a clausal category. For convenience, we in many cases enter just substrings of the labels used for lexvals, such as '`oblDECL`' in the second line, which can be part of both '`__intrObl-oblDECL`' and '`__trObl-oblDECL`', the substring '`-oblDECL`' thereby comprising both of these frame type labels. While the number for clausal lexvals in each category is generally quite close to the total number for that category in NorVal-1952 (shown in the rightmost column), the last two lines illustrate how the number for some non-clausal lexval types in *"The SymbVal Semvalpods"* is radically smaller than their counterpart in NorVal-1952 as a whole.

As can be seen from the third column, nearly 2000 of the lexvals in *"The SymbVal Semvalpods"* are clausal. This is as expected from the definition of the feature `SymbVal`, viz. that the act or state in question has a symbolic function as part of its unfolding. This number is about 1/3 of all the lexvals in "*The SymbVal Semvalpods*" (Table 10 showing mainly only this 1/3), but only 1/5 of all the lexvals in NorVal-1952.

Table 10 serving as a valence profile, the corresponding semantic profile is shown in Table 11, showing the semantic features with which `SymbVal` most typically occurs, again compared with the overall numbers in NorVal-1952 (to be read, for the first line: In *The SymbVal Semvalpods* there are 788 semvalpods whose sempod contains `SymbVal` cooccurring with the feature `Cogn`; in the entire NorVal-1952,

Table 10 Clausal lexvals obtaining in *"The SymbVal Semvalpods"*

Frame type, indicated by category label (or substring of the label) for the type	Reading of label	Number of lexvals with the frame type in "*The SymbVal Semvalpods*"	Number of lexvals with the frame type in NorVal-1952
oblEq	Oblique is a controlled infinitive	623	778
oblDECL	Oblique is a declarative clause	439	515
obDECL	Object is a declarative clause	327	405
oblINTERR	Oblique is an interrogative clause	309	364
obINTERR	Object is an interrogative clause	229	245
obEq	Object is a controlled infinitive	151	165
intrObl-oblN	Oblique is a noun, occurring in an `intrObl` frame	145	987
PRTOF	Oblique with a 'part-of' reading	0	80

there are 937 semvalpods whose sempod contains, possibly among further ones, the feature `Cogn`):

Only the features `Directionslity` and `Loc` from NorVal-1952 are *not* represented among cooccurrences with `SymbVal`. Table 11 in addition shows that `Perceptible, Criterial, PhysBod, Impact, Perform,` and `Procedure` have far fewer appearances in the semvalpods containing `SymbVal` than in the overall NorVal-1952.

The profiles in Tables 10 and 11 give a rather 'holistic' view of the role of the feature `SymbVal`, due to the large number of semvalpods where `SymbVal` is a sempod-member. To get closer to potential profiles of specific verbs, one must intersect the sempods containing `SymbVal` with the sempods containing other features, analogously to how"*The Four Features*" ((17), repeated) was used to arrive at the set of semvalpods "*The 39 semvalpods*" described in Sect. 3 and shown in Appendix 3.

(17)*The Four Features* (repeated from Sect. 3).

`{Impact & Perform & Criterial & PhysBod}`.

The features in (17) when taken separately apply to, respectively, 953, 1401, 391 and 945 verbs (cf. Table 1), and the sempods of their intersections give more pointed views of verbs instantiating them. We now try to establish a corresponding sempod, or sempods, with `SymbVal` as a member, to get a better view of which verbs `SymbVal` typically qualifies.

Table 11 Features cooccurring with `SymbVal` in *The SymbVal Semvalpods*

Feature	Number of cooccurrences with `SymbVal`	Number of occurrences of the feature in NorVal-1952 (from Table 1)
Cogn	788	937
Procedure	711	1148
Perform	555	1401
Response	485	648
Impact	408	953
VerbalAct	283	335
PhysBod	281	945
Domin	268	405
Creation	223	272
Emotion	182	226
Unfold	176	390
Dependence	135	204
Criterial	70	391
MedByPrcpt	68	80
Poss	39	63
PureRel	30	30
Perceptible	25	204
Perceiving	13	26
ImpactSensory	7	14

This is illustrated in Table 12, showing a series of successive intersections of sempods containing `SymbVal` with sempods containing various of the features from Table 11, selected from top and downwards. That means that the features may have a high degree of compatibility with `SymbVal` and thus maybe lead to what may be perceived as a coherent 'class' of verbs (for instance based on `Cogn` and `SymbVal`), but there is no a priori guarantee.

We refer to the lowest line in Table 12 as "*The set of 7*", repeated below:

(27) *The Set of 7*

`{SymbVal & Procedure & Cogn & Perform & Response & Impact & Creation}`.

As seen from the last line, using *The set of 7* as query item, we arrive at a set of 20 semvalpods. These are shown in Appendix 2, called "*The 20 semvalpods*". These semvalpods thus have as their semantic commonality the sempod constituted by the intersection of the respective sempods where `SymbVal`, `Procedure`, `Cogn`, `Perform`, `Response`, `Impact` and `Creation` are members. As a quick excerpt from the leftmost column of "*The 20 semvalpods*", English counterparts to the verbs with these features are the following, each verb thus with a 'foot' in each of the seven features:

Table 12 Sempod schemata containing `SymbVal` in combination with other features

Schema	Instances of semvalpods containing the schema
SymbVal & Procedure & Cogn	544
SymbVal & Procedure & Cogn & Perform	291
SymbVal & Procedure & Cogn & Perform & Response	129
SymbVal & Procedure & Cogn & Perform & Response & Impact	56
SymbVal & Procedure & Cogn & Perform & Response & Impact & Domin	23
SymbVal & Procedure & Cogn & Perform & Response & Impact & Creation	20

(28) Utilize, experiment, legislate, mobilize, recalculate, translate, update, reindex, confirm, adapt, facilitate, underline, highlight, develop, supervise, verify, sanctify, practice

Whether this would meet the expectancy for a 'semantically natural' group or profile is not given, but it illustrates the kind of result one may get with multiple cross-classification into a large inventory of items. There is no lexval common to all these verbs.[13] Among their lexvals, *clausal lexvals* appear in large numbers in *"The 20 semvalpods"*, marked in boldface in Table 13.

As shown in the table, 28 out of 78 lexval instances, i.e., about 35%, are clausal. In comparison, for `Perceiving`, 27 out of 110 lexval instances are clausal (cf. Table 9), i.e., about 25%, and for `ImpactSensory`, 4 out of 88, i.e., less than 5%, are clausal (cf. Table 6). The 236 lexval instances found in '*The 39 semvalpods*' discussed in Sect. 3, analyzed relative to '*The Four Features*', had hardly any instances of clausal lexvals. These observations are summarized in Table 14.

The 1952 verbs of NorVal-1952 represent 1952 semvalpods, which carry altogether 9722 lexvals. The individual features `ImpactSensory` and `Perceiving` represent only 43 of these semvalpods and 198 of the lexvals, thus 0.04 and 0.02%, respectively, of the total NorVal-1952 inventory. Profiles reflecting most of the other features become accessible only when specific sets of them are used for querying, as we have indicated above. The expanse of possibilities thereby to be explored is huge, and may require methodologies well beyond what we have illustrated so far.

[13] 'V__tr' comes close, appearing in all the valpods except the one for *experimentere* 'experiment'. We thus don't have a counterpart to the situation shown in Table 4 in Sect. 3, where the sempod schema "*The Four Features*" combines with the valpod schema "*The Three Lexvals*" to form a 'butterfly' semvalpod with one instantiating verb (*skyte* 'shoot') (or 4 verbs if reducing the three lexvals to two). The number of semantic features in the present case being 7 as opposed to 4 in the case in Table 4, it is not surprising that the intersection potential at the valpod side in the present case is smaller.

Table 13 Profile of lexval types in *"The 20 semvalpods"*

Lexval type	Number of occurrences of the lexval type	Counting as 'clausal'
__tr	19	
__trObl-oblN	9	
__trObl-obRefl-oblN	6	
__tr-obDECL	6	6
__trObl-oblEqObIn	5	5
__tr-obINTERR	5	5
__intrObl-oblN	4	
__trObl-obRefl-oblEqObInf	4	4
__trObl-obRefl-oblDECL	2	2
__trObl-oblDECL	2	2
__intr	2	
__intrObl-oblINTERR	1	1
__ditr-iobRefl	1	
__ditr-iobRefl-obDECL	1	1
__ditr-iobRefl-obINTERR	1	1
__trObl-oblAbsinf	1	1
__trScpr-scPredprtcl	1	
Total	78	28

Table 14 Summaries concerning 'clausal' lexvals

Nickname for set of semvalpods	Location	Number of semvalpods	Proportion of clausal lexvals	Sempod schemata serving as query basis
"The 39 semvalpods"	Appendix 3	39	8/236 (cf. Table 3)	*"The Four Features"*: {Impact & Perform & Criterial & PhysBod}
"The `ImpactSensory` semvalpods"	Table 5	21	4/88 (cf. Table 6)	{ImpactSensory}
"The `Perceiving` semvalpods"	Appendix 1	22	27/110 (cf. Table 8)	{Perceiving}
"The 20 Semvalpods"	Appendix 2	20	28/81 (cf. Table 13)	*"The set of 7"*: {`SymbVal` & `Procedure` & `Cogn` & `Perform` & `Response` & `Impact` & `Creation`}

6 Concluding Remarks

At the outset we stated the proposition *Meaning Determines Valence (MDV)*:

(1) The meanings carried by the verbs of a language determine (within the general limits of the grammar) the verb valence system of the language.

We have approached MDV through exploring the question (3):

(3) Correlation Profiles:
In any given language, are there particular aspects of *verb meanings* which are reflected in aspects of *verb valence*, and vice versa?

NorVal-1952 constitutes already, for the 1952 verbs considered, succinct representations of all lexically attested correspondences between verb meanings and verb valences. What is presently being considered is whether one can state tendencies *over* these correspondences—if one likes, finding *motivated* pairings of meanings with valences. For instance, in the valence correlates to the meaning of 'sensory impact', the circumstance that there is a frequent use of the 'part-of' construction and limited use of clausal constituents is an example of a meaning-valence correspondence, along with the circumstance that the meaning of 'perceiving' supports more constituents of clausal content and syntactic patterns supporting willed action (cf. 5.1). These constructions constitute only 0.02 per cent of the whole NorVal-1952 inventory of lexvals, whereas the figures established around the features of 'symbolic actions' and around acts like 'kicking' cover far more ground. Overall contrasts between these also involve presence vs. absence of clausal constructions but now involving large sets of verbs and interplay between multiple semantic factors. Sections 3 and 5 give illustrations of how one can state partial correspondences between the two domains.

Strengths of the investigation are the large inventory of detailed lexical valence frames, the system of semantic 'meta-features' representing verb semantics, the explicit pairing of valence frames with semantic representations, and the circumstance that valence frames and semantic specifications are represented on a simple common formal format for being paired and queried into. This allows for queries to be made relating to any correlation between a set of semantic features and a set of valence frames.

A formal basis of the system of meta-features is outlined in Sect. 2, relating it to the general formalism of typed feature structures. A pertinent issue pertaining to the notion of 'verb meaning' is that of *subsenses*, for which we provide a representation format in Sect. 4, along with a delineation against sub/supertype relations between meaning representations.

Natural follow-ups to what has been presented include:

a. Exploring possibilities of subjecting the NorVal-1952 structures to more statistical/computational approaches for establishing profiles for the larger domains.
b. Consolidating the meta-feature assignments to the 1952 verbs and extending the system to all the 7500 verbs of Norval, with its 17,500 verb specific valence entries (lexvals).

c. Exploring possible formal connections between the metafeatures (MF) system, including its typed feature structure underpinning, and *semantic/thematic role* systems as have been deployed in many resources. One of these being *NoVRol* (cf. [19]), which is an assignment of roles to all the lexvals of NorVal, an alignment between MFs and the system of roles used in NoVRol could further the assignment mentioned in b.
d. These consolidations and amendments will have to include implementation of subsense representations into the overall system.
e. For the 'daughter' system EngNoVal relative to NorVal-1952, likewise consolidating its meta-feature assignments, and extending the system on to a size like that of NorVal.
f. Orthogonal to the initiatives mentioned, extending the overall architecture of NorVal-1952 to other languages, including languages with far less valence resources but established verb lexicons. This can be done following the strategies used for developing EngNoVal accompanied by language specific studies of contrastive valence, and in turn investigating meaning-valence correspondences with suitably adapted versions of the procedures we have explored here.

Resources

The file '*Sempods-and-Valpods*' representing the full set of semvalpods of NorVal-1952 is located at:

https://github.com/Regdili-NTNU/NorSource/tree/master/NorVal_files.

Included here are also the full NorVal and the file for EngNoVal.

A README file for the resources is about to be included.

Acknowledgements I am grateful to the four reviewers for highly constructive comments.

Disclosure of Interests The author has no competing interests to declare that are relevant to the content of this article.

Appendices

Appendix 1. The semvalpods with sempods containing the feature `Perceiving`

Compared to the semvalpods with sempods containing `ImpactSensory`, far fewer subsenses are here indicated—only three verbs (*kjenne, lukte, smake*) here carry the `SUB`-marking and only these have examples provided. Unlike in Table 5, on the other hand, for these three verbs we include also subsenses not marked for the targeted feature (i.e., in this case, not marked for `Perceiving`) (Table 15).

Appendix 2. The 20 semvalpods"—semvalpods with sempods containing 'The sct of 7'

Table 15 Semvalpods for verbs with 'Perceiving' as member of the sempod

betrakte ('consider')	{Perceiving & Cogn & Procedure & Perform & PhysBod}	{betrakte:V__tr & V__trScpr-scObNrg-scBareinf & V-som__trScpr-scPredprtcl}_3-items	
bivåne ('watch')	{Perceiving & Cogn & Procedure & Perform & PhysBod} &	{bivåne:V__tr & V__tr-obDECL & V__tr-obINTERR}_3-items	
fiksere ('fixate')	{Perceiving & Cogn & Perform}	{fiksere:V__tr & V-på__intrObl-oblDECL & V-på__intrObl-oblEqSuInf & V-på__intrObl-oblINTERR & V-på__intrObl-oblN & V-på__trObl-oblN}_6-items	
glo ('stare')	{Perceiving & Procedure & Unfold & Perform & Criterial & PhysBod}	{glo:V__intr & V-på__intrObl-oblDECL & V-på__intrObl-oblN}_3-items	
hallusinere ('hallucinate')	{Perceiving & Impact & Emotion & SymbVal & PhysBod} &	{hallusinere: V__intr & V__tr & V__tr-obDECL}_3-items	
iaktta ('observe')	{Perceiving & Cogn & Procedure & Response & Unfold & Perform & Domin & SymbVal & PhysBod}	{iaktta:V__tr & V__tr-obDECL & V__tr-obINTERR}_3-items	
kike ('look')	{Perceiving & Unfold & Perform & Criterial & PhysBod}	{kike:V__intr-suDir & V-etter__intrObl-oblN & V-i__intrObl-oblN & V-på__intrObl-oblDECL & V-på__intrObl-oblN}_5-items	
kikke ('look')	{Perceiving & Unfold & Perform & Criterial & PhysBod}	{kikke:V__intr-suDir & V-etter__intrObl-oblN & V-i__intrObl-oblN & V-på__intrObl-oblDECL & V-på__intrObl-oblN}_5-items	

(continued)

Table 15 (continued)

betrakte ('consider')	{Perceiving & Cogn & Procedure & Perform & PhysBod}	{betrakte:V__tr & V__trScpr-scObNrg-scBareinf & V-som__trScpr-scPredprtcl}_3-items	
kjenne_SUB_EXP ('feel')	{ImpactSensory & MedByPrcept & Cogn & Impact & Response & SymbVal & PhysBod}	{kjenne:V__trScpr-obRefl-scObNrg-scBareinf & V__trScpr-scObNrg-scBareinf}_2-items	hun kjenner seg stivne & jeg kjenner det stivne &
kjenne_SUB_AG ('feel')	{Perceiving & Response & Perform & PhysBod}	{kjenne:V-på__intrObl-oblN}_1-items	hun kjenner på pakken &
kjenne_SUB_COG ('feel, know')	{MedByPrcept & Cogn & Response & SymbVal & PhysBod}	{kjenne:V-til__intrObl-oblDECL & V-til__intrObl-oblINTERR & V-til__intrObl-oblN & V__tr & V__tr-obDECL & V__tr-obINTERR & V__trScpr-obRefl-scObNrg-scPred & V__trScpr-obRefl-scObNrg-scBareinf & V__trScpr-scObNrg-scBareinf & V-igjen__trPrtcl & V-igjen__trPrtcl-obINTERR & V-igjen__trPrtcl-obRefl & V-som__trScpr-obRefl-scPredprtcl & V-som__trScpr-scPredprtcl}_14-items	hun kjenner til at vi kommer & hun kjenner til hvem som kommer & vi kjenner til problemet & vi kjenner komitéformannen & vi kjenner at det blir varmere & han kjenner hvor smerten oppstår & hun kjenner seg trygg & hun kjenner seg stivne & jeg kjenner det stivne & hun kjenner igjen ansiktet & de kjenner igjen hva som er i ferd med å skje & hun kjenner seg igjen & hun kjenner seg som en vinner & vi kjenner dem som konkurrenter
lukte_SUB_AG ('smell')	{Perceiving & ImpactSensory & Cogn & Response & Perform & Criterial & SymbVal & PhysBod}	{lukte:V-på__intrObl-oblN}_1-items	vi lukter på blomstene &

(continued)

Table 15 (continued)

betrakte ('consider')	{Perceiving & Cogn & Procedure & Perform & PhysBod}	{betrakte:V__tr & V__trScpr-scObNrg-scBareinf & V-som__trScpr-scPredprtcl}_ 3-items	
lukte_SUB_ COG ('smell')	{Perceiving & ImpactSensory & Cogn & Response & Perform & Criterial & SymbVal & PhysBod}	{lukte:V__tr & V__tr-obDECL & V__tr-obINTERR &}_3-items	hun lukter blomstene & hun lukter at noen har vært der & hunden lukter om det har vært katter der &
lytte ('listen')	{Perceiving & Procedure & Response & Perform & Criterial & PhysBod}	{lytte:V__intr & V-på__ intrObl-oblN & V-til__ intrObl-oblN}_3-items	
observere ('observe')	{Perceiving & Cogn & Procedure & Perform}	{observere:V__tr-obINTERR & V__tr & V__tr-obDECL & V__ trScpr-scObNrg-scBareinf}_ 4-items	
persipere ('perceive')	{Perceiving & Response & Perform & SymbVal & PhysBod}	{persipere:V__tr & V__ tr-obDECL & V__ tr-obINTERR}_3-items	
sanse ('sense')	{Perceiving & SymbVal & PhysBod}	{sanse:V__tr & V__tr-obDECL & V__tr-obINTERR & V__ tr-obRefl}_4-items	
skimte ('glimpse')	{Perceiving & Cogn & Perform & Criterial & SymbVal & PhysBod}	{skimte:V__tr & V__tr-obDECL & V__tr-obINTERR}_3-items	
skotte ('peek')	{Dir & Perceiving & Perform & Criterial & PhysBod}	{skotte:V__intr & V__intr-suDir & V-på__intrObl-oblN}_3-items	

(continued)

Table 15 (continued)

betrakte ('consider')	{Perceiving & Cogn & Procedure & Perform & PhysBod}	{betrakte:V__tr & V__trScpr-scObNrg-scBareinf & V-som__trScpr-scPredprtcl}_3-items	
skule ('scowl')	{Dir & Perceiving & Emotion & Perform & Criterial & PhysBod}	{skule:V__intr & V-mot__intrObl-oblN & V-på__intrObl-oblN}_3-items	
smake_SUB_AG ('taste')	{Perceiving & ImpactSensory &Response & Perform & Criterial & SymbVal & PhysBod}	{smake:V-på__intrObl-oblN}_1-items	vi smaker på suppen

Table 16 shows semvalpods with sempods containing the set `{Cogn & Impact & Procedure & Creation & Response & Perform & SymbVal}`, called *The 7 Semvalpods* in Sect. 5.2.

Appendix 3. The 39 Semvalpods

Semvalpods for *FourFeatures* (short version, based on where the features are contiguous in their sempod, amounting to 39 semvalpods. When the features are not only contiguous, the number of semvalpods is 182). This is a text version of a semvalpod, cf. (15a).

bite & ('bite') & {Sustain & Perceptible & Impact & Perform & Criterial & PhysBod} & VALPOD: {bite:V__intr & V-i__intrObl-oblN & V-i__trScpr-scPPrefl & V__tr & V-av__trPrtcl & V-sammen__trPrtcl & V-fast__trPrtcl-obRefl}_7-items

brekke & ('break') & {Sustain & Perceptible & Impact & Perform & Criterial & PhysBod} & VALPOD: {brekke:V__intr & V-av__intrPrtcl & V-av-på__intrPrtclObl-oblPRTOFsu & V__tr & V__tr-obRefl & V-av__trPrtcl & V-av-på__trPrtclObl-oblPRTOFsu & V-opp__trPrtcl}_8-items

bøye & ('bend') & {Sustain & Directionality & Perceptible & Dependence & Impact & Perform & Criterial & PhysBod} & VALPOD: {bøye:V-av__intrPrtcl & V-av-for__intrPrtclObl-oblN & V-på__intrObl-oblN & V__tr & V__tr-obDir & V__tr-obRefl & V-for__trObl-obRefl-oblN & V-av__trPrtcl}_8-items

glefse & ('gulp, bite') & {Impact & Perform & Criterial & PhysBod} & VALPOD: {glefse:V__intr & V-etter__intrObl-oblN & V-i__trScpr-scPPrefl}_3-items

gripe & ('grasp') & {Init & Cogn & Impact & Perform & Criterial & PhysBod} & VALPOD: {gripe:V__tr & V-etter__intrObl-oblN & V-inn__intrPrtcl & V-inn-i__intrPrtclObl-oblN & V-inn-overfor__intrPrtclObl-oblN}_5-items

Table 16 "The 20 semvalpods"

Verb	Sempod	Valpod
benytte ('utilize')	{Cogn & Impact & Procedure & Creation & Response & Perform & SymbVal}	{benytte:V__tr & V-av__trObl-obRefl-oblN & V-av__trObl-obRefl-oblDECL}_3-items
eksperimentere ('experiment')	{Cogn & Impact & Procedure & Creation & Response & Perform & SymbVal}	{eksperimentere:V__intr & V__trPath-obRefl-obDir & V-med__intrObl-oblAbsinf & V-med__intrObl-oblN}_4-items
lovfeste ('legislate')	{Cogn & Impact & Procedure & Creation & Response & Perform & Domin & SymbVal}	{lovfeste:V__tr & V__tr-obDECL & V__tr-obINTERR}_3-items
mobilisere ('mobilize')	{Cogn & Impact & Procedure & Creation & Response & Perform & Domin & SymbVal & PhysBod}	{mobilisere:V__tr & V-til__trObl-oblEqObInf & V-til__trObl-oblN}_3-items
omregne ('recalculate')	{Cogn & Impact & Procedure & Creation & Response & Perform & SymbVal}	{omregne:V__tr & V-fra__intrObl-oblN & V-til__intrObl-oblN}_3-items
omsette-1 ('translate')	{Cogn & Impact & Procedure & Creation & Response & Perform & SymbVal}	{omsette-1:V__tr & V-fra__trObl-oblN & V-til__trObl-oblN}_3-items
oppdatere ('update')	{Cogn & Impact & Procedure & Creation & Response & Perform & SymbVal}	{oppdatere:V__tr & V__tr-obRefl & V-på__trObl-oblEqObInf & V-på__trObl-oblINTERR & V-på__trObl-oblN & V-på__trObl-obRefl-oblEqObInf & V-på__trObl-obRefl-oblINTERR & V-på__trObl-obRefl-oblN}_8-items
oppjustere ('adjust up')	{Cogn & Impact & Procedure & Creation & Response & Perform & Domin & SymbVal}	{oppjustere:V__tr & V-til__trObl-oblEqObInf & V-til__trObl-oblN}_3-items
reindeksere ('re-index')	{Cogn & Impact & Procedure & Creation & Response & Perform & SymbVal}	{reindeksere:V__tr & V-for__intr-oblINTERR & V-for__intrObl-oblN}_3-items

(continued)

Table 16 (continued)

Verb	Sempod	Valpod
stadfeste ('confirm')	{Cogn & Impact & Procedure & Creation & Response & Perform & Domin & SymbVal}	{stadfeste:V__tr & V__tr-obDECL & V__tr-obINTERR}_3-items
tilpasse ('adapt')	{Cogn & Dependence & Impact & Procedure & Creation & Response & Perform & Domin & SymbVal}	{tilpasse:V__ditr-iobRefl & V__ditr-iobRefl-obDECL & V__ditr-iobRefl-obINTERR & V__tr & V-til__trObl-oblDECL & V-til__trObl-oblN & V-til__trObl-obRefl-oblDECL & V-til__trObl-obRefl-oblN & V-til__trObl-obRefl-oblINTERR & V-til__trObl-obRefl-oblINTERR}_10-items
tilrettelegge ('facilitate')	{Cogn & Impact & Procedure & Creation & Response & Perform & Domin & SymbVal}	{tilrettelegge:V__tr & V-for__trObl-oblAbsinf & V-for__trObl-oblDECL & V-for__trObl-oblN}_4-items
understreke ('underline')	{Cogn & Impact & Procedure & Creation & Response & Perform & SymbVal}	{understreke:V__tr & V__tr-obDECL & V__tr-obINTERR}_3-items
understryke ('underline')	{Cogn & Impact & Procedure & Creation & Response & Perform & SymbVal}	{understryke:V__tr & V__tr-obDECL & V__tr-obINTERR}_3-items
utheve ('highlight')	{Cogn & Impact & Procedure & Creation & Response & Perform & SymbVal}	{utheve:V__tr & V__tr-obDECL & V__tr-obINTERR}_3-items
utvikle ('develop')	{Cogn & Impact & Procedure & Creation & Response & Perform & SymbVal}	{utvikle:V__tr & V__tr-obRefl & V-til__trObl-obRefl-oblEqObInf & V-til__trObl-obRefl-oblN}_4-items
veilede ('supervise')	{Cogn & Impact & Procedure & Creation & Response & Perform & Domin & SymbVal}	{veilede:V__tr & V-i__trObl-oblEqObInf & V-i__trObl-oblINTERR & V-i__trObl-oblN}_4-items
verifisere ('verify')	{Cogn & Impact & Procedure & Creation & Response & Perform & SymbVal}	{verifisere:V__tr & V__tr-obDECL & V__tr-obINTERR}_3-items

(continued)

Table 16 (continued)

Verb	Sempod	Valpod
vigsle ('sanctify')	{Cogn & Dependence & Impact & Procedure & Creation & Response & Perform & Domin & SymbVal}	{vigsle:V__tr & V-til__trObl-oblEqObInf & V-som__trScpr-scPredprtcl}_3-items
øve ('practice')	{Cogn & Impact & Procedure & Creation & Response & Unfold & Perform & SymbVal}	{øve:V__intr & V-i__trObl-oblN & V__tr & V__tr-obRefl & V-på__trObl-obRefl-oblEqObInf & V-på__trObl-obRefl-oblN & V-til__trobl-obRefl-oblEqObInf & V-til__trObl-obRefl-oblN}_8-items

hakke & ('hack') & {Sustain & Perceptible & Impact & Perform & Criterial & PhysBod} & VALPOD: {hakke:V__intr & V__tr & V-ned__trPrtcl & V-opp__trPrtcl & V-på__intrObl-oblN}_5-items

hogge & ('chop') & {Sustain & Perceptible & Cogn & Impact & Perform & Criterial & PhysBod} & VALPOD: {hogge:V__tr & V__trObl-oblPRTOFob & V__trObl-obRefl-oblPRTOFob & V-av__trPrtcl & V-ned__trPrtcl & V-over__trPrtcl & V-til__trPrtcl}_7-items

hoppe & ('jump') & {Directionality & Impact & Perform & Criterial & PhysBod} & VALPOD: {hoppe:V__intr-suDir & V__trScpr-scObNrgCsd-scPred & V__trScpr-obRefl-scObNrgCsd-scPred & V__tr-suDir-obLengthunit}_4-items

jumpe & ('jump') & {Directionality & Impact & Perform & Criterial & PhysBod} & VALPOD: {jumpe:V__intr & V__intr-suDir & V-i__intrObl-oblN}_3-items

kaste & ('throw') & {Sustain & Directionality & Perceptible & Impact & Perform & Criterial & PhysBod} & VALPOD: {kaste:V__ditrObl-oblPRTOFiob & V-på__intrObl-oblPRTOFsu & V__tr-obDir & V__tr-obRefl & V-opp__intrPrtcl}_5-items

kile & ('tickle') & {ImpactSensory & Impact & Perform & Criterial & PhysBod} & VALPOD: {kile:V__impersObl-oblN & V__tr & V__trObl-oblPRTOFob & V__trObl-obRefl-oblPRTOFob & V-inn__trPrtcl & V-inn__trPrtcl-obRefl}_6-items

klippe & ('cut, clip') & {Sustain & Perceptible & Impact & Perform & Criterial & PhysBod} & VALPOD: {klippe:V__tr & V__tr-obRefl & V-av__trPrtcl & V-bort__trPrtcl & V-ut__trPrtcl}_5-items

klubbe & ('club') & {Sustain & Perceptible & Impact & Perform & Criterial & PhysBod} & VALPOD: {klubbe:V__intr & V__tr & V__trObl-oblPRTOFob & V-gjennom__trPrtcl & V-ned__trPrtcl}_5-items

kludre & ('meddle') & {Imped & Impact & Perform & Criterial & PhysBod} & VALPOD: {kludre:V__intr & V-med__intrObl-oblN & V-til__trPrtcl}_3-items

knyte & ('knit') & {Sustain & Perceptible & Impact & Perform & Criterial & PhysBod} & VALPOD: {knyte:V__tr & V__tr-obRefl & V-i__trImpersObl-obRefl & V-i-på__trImpersObl2-obRefl-obl1PRTOFobl2 & V-igjen__trPrtcl & V-opp__trPrtcl & V-sammen__trPrtcl & V-sammen-i__trImpersPrtclObl-obRefl & V-sammen-i-på__trImpersPrtclObl2-obRefl-obl1PRTOFobl2}_9-items

knytte & ('knit') & {Sustain & Impact & Perform & Criterial & PhysBod} & VALPOD: {knytte:V__tr & V__tr-obRefl & V-i__trImpersObl-obRefl & V-i-på__trImpersObl2-obRefl-obl1PRTOFobl2 & V-igjen__trPrtcl & V-opp__trPrtcl & V-sammen__trPrtcl & V-sammen-i__trImpersPrtclObl-obRefl & V-sammen-i-på__trImpersPrtclObl2-obRefl-obl1PRTOFobl2}_9-items.

krysse-1 & ('check') & {Sustain & Cogn & Impact & Perform & Criterial & PhysBod} & VALPOD: {krysse-1:V-av__intrObl-oblN & V-av__trPrtcl & V-ut__trPrtcl}_3-items

lange & ('provide, hand') & {Sustain & Directionality & Impact & Perform & Criterial & PhysBod} & VALPOD: {lange:V-i__trObl-oblRefl & V-i-av__intrObl2-obl1Refl & V-til__intrObl-oblN & V-ut__intrPrtcl & V-til__trObl-oblN}_5-items

lempe & ('throw') & {Directionality & Impact & Perform & Criterial & PhysBod} & VALPOD: {lempe:V-på__intrObl-oblDECL & V-på__intrObl-oblN & V__tr-obDir & V__tr-obRefl & V-etter__trObl-obRefl-oblN}_5-items

løse & ('solve') & {Sustain & Dependence & Impact & Perform & Criterial & PhysBod} & VALPOD: {løse:V__tr & V__tr-obRefl & V-fra__trObl-oblN & V-opp__trPrtcl & V-opp__trPrtcl-obRefl}_5-items

nappe & ('pinch') & {Sustain & Directionality & Perceptible & Impact & Perform & Criterial & PhysBod} & VALPOD: {nappe:V__intr & V__tr & V-av__trObl-oblN & V-bort__trPrtcl & V-ut__trPrtcl & V-fra__trObl-oblN & V-til__trScpr-scPPrefl}_7-items

prikke & ('poke') & {Sustain & Cogn & Impact & Perform & Criterial & PhysBod} & VALPOD: {prikke:V-i__impersObl-oblN & V__intrObl-oblLoc & V__tr & V__trObl-oblPRTOFob & V-inn__trPrtcl}_5-items

puffe & ('push') & {Sustain & Perceptible & Impact & Perform & Criterial & PhysBod} & VALPOD: {puffe:V__intr & V__tr-obDir & V-til intrObl-oblN}_3-items

rikke & ('move') & {Directionality & Impact & Perform & Criterial & PhysBod} & VALPOD: {rikke:V-på__intrObl-oblN & V-på__trObl-oblRefl & V__tr-obDir & V__tr-obRefl-obDir}_4-items

røre-1 & ('touch, move') & {Perceptible & Impact & Perform & Criterial & PhysBod} & VALPOD: {røre-1:V-på__intrObl-oblN & V-på__intrObl-oblRefl & V-ved__intrObl-oblN & V__tr & V__tr-obRefl}_5-items

sjaske & ('mess') & {Imped & Impact & Perform & Criterial & PhysBod} & VALPOD: {sjaske:V__intr & V-til__trPrtcl & V-til__trPrtcl-obRefl}_3-items

skyte & ('shoot') & {Sustain & Directionality & Cogn & Impact & Perform & Criterial & PhysBod} & VALPOD: {skyte:V__intr-suDir & V-på__intrObl-oblN & V-av__intrPrtcl & V__tr-obDir & V__tr-obRefl & V__trObl-oblPRTOFob & V__trObl-obRefl-oblPRTOFob & V-på__trObl-oblN & V-inn__trPrtcl-obDECL &

V-inn__trPrtcl-obINTERR & V-til__trPrtcl-obDECL & V__trScpr-scObCsd}_12-items

sleike & ('lick') & {Sustain & Perceptible & Impact & Perform & Criterial & PhysBod} & VALPOD: {sleike:V-av__intrObl-oblN & V-på__intrObl-oblN & V__tr & V__trObl-oblPRTOFob & V__trObl-obRefl-oblPRTOFob & V-i__trScpr-scPPrefl}_6-items

smekke & ('smack') & {Sustain & Perceptible & Impact & Perform & Criterial & PhysBod} & VALPOD: {smekke:V__intr & V__trObl-oblPRTOFob & V-igjen__trPrtcl}_3-items

springe & ('run, spring') & {Sustain & Directionality & Impact & Perform & Criterial & PhysBod} & VALPOD: {springe:V__intr & V__intrPresntDir & V__intr-suDir & V-ut__intrPrtcl}_4-items

spytte & ('spit') & {Sustain & Directionality & Perceptible & Impact & Perform & Criterial & PhysBod} & VALPOD: {spytte:V__intr & V-på__intrObl-oblN & V-ut__intrPrtcl & V__tr & V-på__trObl-oblN & V-ut__trPrtcl}_6-items

stampe & ('stomp') & {Sustain & Directionality & Impact & Perform & Criterial & PhysBod} & VALPOD: {stampe:V__intr & V__intrObl-oblLoc & V__tr & V__trScpr-scObCsd}_4-items

stikke-1 & ('sting') & {Sustain & ImpactSensory & Perceptible & Impact & Perform & Criterial & PhysBod} & VALPOD: {stikke-1:V__intr & V-med__intrObl-oblN & V__tr-obDir & V__trObl-obRefl-oblPRTOFob & V-med__trObl-oblN & V-ut__trPrtcl-obRefl}_5-items.

stille & ('pose') & {Sustain & Impact & Perform & Criterial & PhysBod} & VALPOD: {stille:V__ditr & V-som__intrLghtScpr-scSuNrg-scPredprtcl & V-mot__intrObl-oblN & V-opp__intrPrtcl & V-opp-mot__intrPrtclObl-oblAbsinf & V-opp-mot__intrPrtclObl-oblN & V__tr & V__tr-obDir & V__trObl-obRefl-oblLoc & V__trPath-obRefl-obDir & V-inn__trPrtcl & V__trScpr-scObLoc & V__trScpr-obRefl-scPred}_13-items

stramme & ('tighten') & {Sustain & Impact & Perform & Criterial & PhysBod} & VALPOD: {stramme:V__intr & V__tr & V-igjen__trPrtcl & V-inn__trPrtcl & V-opp__trPrtcl & V-opp__trPrtcl-obRefl}_6-items

stø & ('balance') & {Sustain & Cogn & Dependence & Impact & Perform & Criterial & PhysBod} & VALPOD: {stø:V__intr & V__tr-obDir & V-på__trObl-obRefl-oblDECL & V-på__trObl-obRefl-oblEqObInf & V-på__trObl-obRefl-oblINTERR & V-på__trObl-obRefl-oblN}_6-items

støe & ('balance') & {Sustain & Cogn & Dependence & Impact & Perform & Criterial & PhysBod} & VALPOD: {støe:V__intr & V__tr-obDir & V-på__trObl-obRefl-oblDECL & V-på__trObl-obRefl-oblEqObInf & V-på__trObl-obRefl-oblINTERR & V-på__trObl-obRefl-oblN}_6-items

søle & ('spill, spoil') & {Imped & Impact & Perform & Criterial & PhysBod} & VALPOD: {søle:V__intr & V-med__intrObl-oblN & V-på__intrObl-oblN & V__tr & V-bort__trPrtcl & V-til__trPrtcl & V-til__trPrtcl-obRefl}_7-items

tråkke & ('step, tramp') & {Sustain & Impact & Perform & Criterial & PhysBod} & VALPOD: {tråkke:V__intr & V__intr-suDir & V-i__intrObl-oblN & V-på__intrObl-oblN & V-til__intrPrtcl & V__tr & V__trObl-oblPRTOFob}_7-items

References

1. Hellan L (2024) Meaning features and verb valence. In: International symposium on distributed computing and artificial intelligence. pp 163-172. Cham: Springer Nature Switzerland
2. Hellan L (2022) Natural language processing in artificial intelligence-NLPinAI 2021. Springer
3. Hellan L (2023) Logic and algorithms in computational linguistics 2021. Springer Nature
4. Hellan L (2023) Logic and algorithms in computational linguistics 2021. Springer Nature
5. Hellan L (2025) Meaning features and verb valence. In: Distributed computing and artificial intelligence, Special Sessions I, 21st International Conference (DCAI 2024) lecture Notes in Networks and Systems (LNNS). Publisher, Springer Cham.
6. Hellan L (2023) A unified cluster of valence resources. In: Logic and algorithms in computational linguistics 2021 Springer Nature. ISBN 978–3–031–21780–7
7. Levin B (1994) English verb classes and alternations. University of Chicago Press, Chicago
8. Korhonen A, Briscoe T (2004) Extended lexical-semantic classification of english verbs. In: Proceedings of the HLT/naacl workshop on computational lexical semantics, Boston, MA
9. Malchukov AL, Comrie B (eds) (2015) Valency classes in the world's languages. Mouton De Gruyter, Berlin
10. Hellan L (2022) A valence catalogue for Norwegian. In: Loukanova R (ed) Natural language processing in artificial intelligence-NLPinAI 2021. Springer. URL: https://doi.org/10.18710/8U3L2U
11. Hellan L (2023) A unified cluster of valence resources. In: Logic and algorithms in computational linguistics 2021 (LACompLing2021). Springer Nature 2023 ISBN 978–3–031–21780–7
12. Hellan L (Ms 1) Porting a Norwegian valence resource to English.
13. Hellan L (2025) Meaning features and verb valence. In: Distributed computing and artificial intelligence, Special Sessions I, 21st International Conference (DCAI 2024). Part of the book series: Lecture Notes in Networks and Systems (LNNS). Publisher: Springer Cham
14. Carpenter B (1992) The logic of typed feature structures. Cambridge University Press, Cambridge
15. Copestake A (2002) Implementing typed feature structure grammars. CSLI Publications, Stanford
16. Pollard C, Sag IA (1994) Head-driven phrase structure grammar. Chicago University Press, Chicago
17. Hellan L (2019) Construction-based compositional grammar. J Logic Lang Inform URL: https://doi.org/10.1007/s10849-019-09284-5
18. Hellan L (Ms 2) Extending a system for verb semantics.
19. Torgersen HAT, Ravnanger EØ, Hellan L, Haug DTT (2024) NoVRol: a semantic role lexicon of Norwegian verbs. In: Proceedings of the 2024 joint international conference on computational linguistics, language resources and evaluation (LREC-COLING 2024). European Language Resources Association 2024 ISBN 978–2–493814–10–4
20. Hellan L (2022) A valence catalogue for norwegian. In: Loukanova R (ed) natural language processing in artificial intelligence-NLPinAI 2021. Springer. https://doi.org/10.18710/8U3L2U